A Rooster Named Alice
Stories from Riverbridge Farm

O. Ross McIntyre

Makardua, Lyme NH

"It occurred to me that if we sold this place and bought a real farm that I would get more exercise." —O Ross McIntyre 1968

Table of Contents

Stories from Riverbridge Farm

ISBN: 979-8-9879461-07 (Hardcover edition)
ISBN: 979-8-9879461-1-4 (Paperback edition)
ISBN: 979-8-9879461-2-1 (Ebook edition)

Cover photo by O.Ross McIntyre edited by Richard Granger
Drone photo by Bill Hudenko
All other photos by Jean and O.Ross McIntyre
Farm Map by Kate Emlen
Watercolor by Anne Mellor

Excerpt from Complete Poems of Robert Frost, The Tuft of Flowers
Henry Holt and Company, New York (Permission requested)

Cover Photo: The rooster on the cover is not Alice. He is a stand-in.
Our sociopathic dog, Vincent, ate Alice and I hadn't yet taken
his picture. This rooster is more colorful than Alice who lacked
the fishnet stocking pattern over his breast. David Caffry of Lyme
let me photograph this wonderful bird as he walked amongst
his flock. Perhaps agitated by my presence, the rooster stood on
one foot in the snow, the other leg partly hidden in his feathers, cocked and ready to kick. The sharp spur on the cocked leg
is covered, but at the ready to impale as an add-on to a kick.

Printed in the United States of America

Dedication

To
those who have ever lifted a hay bale, turned a separator crank,
cleaned a chicken, sheared a sheep, lost a calf, worried about
a spring freeze-up, hauled water, beat out a fire,
pulled a weed, carried a hundred pounds,
so the rest of us could eat.

Preface

In 1969 my wife, Jean, and I bought a house and farm on the Connecticut River in Lyme, New Hampshire. We moved our belongings and three children into the house where, for the next eleven years, we enjoyed a life on which this book is based. Together we tended cattle, sheep, and horses. We cared for them at birth and sometimes at death. We collected eggs, made thousands of bales of hay, and fixed our farm equipment when it broke. We learned about weather, and how to use each of the four seasons to best advantage. We also learned to recognize the sweet smell of corn pollen in July, follow the path of the plow where the worms wiggle in the newly turned furrow, and watch coyotes pick mice from the new-mown hay.

During this same time I was responsible for the development of the Cancer Center at the Dartmouth-Hitchcock Medical Center. That was more than a full-time job. The farm was my exercise program, my way to dissipate aggression, my way to avoid losing my way.

In 1981 we had to simplify our lives. So we sold the house and some of the land, shut down most of the farming activities, and hunkered down in a new house for the long-haul that led to the National Cancer Institute award of Comprehensive status for the Norris Cotton Cancer Center. Those wonderful eleven years, however, taught all of the family much about how we fit into the earth's scheme.

Since those years, the "back to nature" culture has withered a bit, and youngsters are increasingly drawn to city lights, instead of ponies, calves and lambs. As we look back in our country's history, we find that the theme covered here has fluctuated in popularity over the decades, sometimes stronger and sometimes weaker, often influenced by changes in the national economy. Sometimes the enthusiasm for a "return to nature" seems to be catalyzed by a single book, or a couple of articulate speakers, or even a catalogue. With the onset of the Covid pandemic, some families sought refuge in the countryside. Despite the changes in popularity however, the land has always beckoned to some adults and children and I hope they have the chance that we had to partake of it. This book is for them, and for those in the city who wish only to dream of such a life. Finally it is for those who have had that life and now, years later, find it possible to reminisce on the best years of their lives.

Chapter 1

Walking the Boundary Without Being Shot

O ne evening in June of 1969, Mike Smith called to inform me that his neighbor, Marcel Sessler, had decided to sell his farm. This was a property that abutted the Smith's. Mike knew that my wife, Jean, and I had been looking for a place where we could have a horse, and here was an opportunity that seemed promising. We dropped our usual evening tasks and hurried to the Smith's place where Mike and his wife, also named Jean, had already spread aerial photographs of that corner of Lyme on their kitchen table by the time we arrived.

"Bob MacDonnell, Sessler's real estate agent, called late this afternoon," said Mike. "He thinks it is a perfect place for your horse." He also emphasized: "Sessler is a peculiar guy who can easily be upset during negotiations. If we lose track of that fact, no deals are possible."

I replied, "So we are dealing with a guy who is cantankerous?

"Right." Mike said. "He is a nut-case. Won't let anyone on his land because they might be carrying a cancer virus. The old-timers refer to him as *The Squire!*"

In addition to the aerial photos, MacDonnell had furnished Mike with a copy of a map of the Sessler land prepared in 1825. As we puzzled over the photos trying to apply some of the landmarks

3

seen in the 1825 map, it was clear that we had a long way to go in learning the bounds of the Sessler place.

Jean and I had visited two other places for sale in Lyme that we felt were of interest. The owners of these properties had walked the boundaries with us uphill, downhill through dense woods, in each case eager to show off what they had for sale. It was clear that this was not going to happen with the Sessler farm. If we wished to examine it we would have to do it without the owner's permission and on our own.

MacDonnell had told Mike that the Squire was attending his class reunion at Cornell for several days and that if we wanted to see the property, this would be a good time to do it. So the next morning, Mike and I set out to walk the boundary line. Early in the morning, we walked from the Smith property into a dense stand of hemlock trees on the Sessler property. The trees were packed so tightly that it was hard to see the red paint blazes on each tree that marked the boundary. From time to time we could spot the rusty barbed wire of the old boundary fence. The wire was so old it broke when we bent it.

About an hour after starting, we came to an old stone wall and continued along the wire, following the blazes on trees that rose up the hill ahead. It was tough going. Loose pine needles lay on sloping rocks. Mike was ahead, moving with purpose, his Vibrams biting into the slippery duff. The chain tread on my Bean boots had been worn away many miles ago and I took twice as many steps as he to go the same distance. Farther back, when we were in the swamp, he sucked mud with his leather boots and I had congratulated myself on my choice of footwear. Now, he could do the same.

Half way up the hill we paused to catch our breath and to swat the flies that swarmed around us. From the east the church clock struck the hour. The sound of the bell came through the mile-and-a-half of woods between us and the village of Lyme as clearly as if we had been standing in the cemetery behind the church. It was a hot summer day in New Hampshire and with high humidity carrying the sound. The bell-note predicted rain before the day was up.

I am handicapped by my Nebraska childhood when it comes to finding my way in deep woods. The grid of north-south, east-west boundary lines and roads found there has conditioned my mind to force every property into a plan with right angle corners. This is seldom the case in New England. Soon things don't add up and I get confused.

We had been following the boundary up hill and down, through the shade of the dense hemlock into a swamp, and now up the steep hill. The line we followed had several corners and by now I could not have placed more than a couple of them accurately on an aerial photograph. Mike is far better at this sort of thing than I am. He said, "I know right where we are. All these ledges run northwest to southeast." We were standing on a carpet of pine needles beneath some weevil-infested white pines whose roots were trying to dissolve the solid rock of the ledge beneath us. I guess he was right; the ledge appeared to run northwest to southeast.

Now we were headed north along the side of a west-facing hill. The old barbed wire was hard to find here. We dropped into a notch where a spring-fed trickle ran to the west. The black mud on its soft bottom came nearly over my boots. The blazes now traced a curving line to the west. A curved line? Surveyors usually make straight lines. The wire was gone. We went back to find it and then followed it until it disappeared at the beginning of the curve. The blazes went on down the hill to a corner, then turned acutely back upon themselves.

We squeezed between a crack in a rock outcropping and found some wire. What was going on here? Probably the person putting up the fence had decided to follow the easiest route to fence rather than the property line. This happened when an acre of woods was of no particular value.

"We are getting close to the house," Mike said, as we could hear the traffic on River Road below us. "Better not get too close.

If MacDonnell is wrong about the day the Squire returns home he should not see us."

We left the line so that we would be further from the road, and went straight north until we came out on an old logging road. The hot sun baked a springing carpet of pine needles. A deer fly or two appeared and buzzed into our hair. I swatted, keeping up the pressure of my hand on my head until I felt the pop that signified death. (There is no point in swatting a tough deer fly only to let it return for a second chance.)

The logging road dropped down, made a sharp turn to the north and crossed a gully holding a small brook. Beyond, the trees opened into a pasture. We walked on into the pasture, climbing along a woven wire fence that formed the southwestern boundary of the field. Most of the wire lay on the ground, the posts having rotted off. "Better get ready to duck," said Mike who was already crouching. "He might have come home early."

We climbed toward the place where the track went through an opening in the fallen fence – a place where there had once been a gate. We could now look down the hill to the west. Below us, a pasture with rock outcroppings fell away for a hundred yards to a high stone wall. Beyond that, timothy heads and Indian paintbrush waved at us from a depleted hay field. We could see the brook we had just crossed. It had dropped down to the west and now turned north separating the hay field from a large level field that lay beyond. Surrounded by this field were barns and a house that fronted on River Road. Across the road a large meadow ran down to the bank of the Connecticut River.

We surveyed the farmhouse from our crouched position. Mike spoke in a whisper, " I don't see any sign of life down there but he could have come home early. He is unpredictable."

"What would happen if he sees us?"

"Well, he will shoot at us."

Chapter 2

The Squire

Well, he might have shot us. A few years earlier two couples had arrived by motor boat for a leisurely afternoon picnic at the south end of the Squire's riverside meadow and spread their tablecloth at the edge of the recently mowed field. As they were settling down for sandwiches and a bit of wine, the "whir-kerchunk" of a bullet flying over their heads and striking the water was followed by a rifle report from the direction of the house. They swept up their tablecloth jumbling the sandwiches, wine, and dessert, as several more bullets went overhead. The motor started, the bullets ceased, and the escape was uneventful except for the involuntary trembling of hands and feet and the sodden tablecloth in the bottom of the boat.

After getting to know the Squire I could imagine how his encounter with the State Trooper had gone. "Sure, I was shooting," the Squire told the officer who found him sitting in the shade of the big maple tree next to the driveway. He rose, his back to the asparagus patch. "Those damn woodchucks are making a mess down there in the mowing next to the River! Let me show you where."

"Look here, Mr. Squire," the officer stretched to his full height and tipped up the rear of his wide-brimmed hat to diminish the space between his eyes and the hat brim, "You know damn well you were shooting at those people having a picnic on your river bank.

7

We have had calls about you before this, and now you have done it again. One of those people you shot at is a trustee of the hospital, the other is the best doc in the Valley."

"No sir. I was shooting at woodchucks. They are right down there on the riverbank. There weren't any people."

The officer looked down at the elderly and stooped figure through thick glasses into the Squire's gray eyes. "Mr. Squire, I take this sort of thing very seriously." (No one had actually seen the Squire fire the rifle.) "I'm not going to arrest you this time, but if I ever have to come back here again, you will be in serious trouble."

A drone's eye view of the River Meadow with River Road and the Riverbridge Farmhouse in the mid distance. The picnic that was interrupted by gunshots was held a few feet up from the Connecticut River at the right end of the field.

My imagination continues: As the trooper's car pulled away, the Squire went through the garage into the mud room where the gun rack was screwed to the north wall of the room. He took down the Hornet 22, removed the empty clip and proceeded to swab the barrel with a small square of flannel on a brass rod. He then reloaded the clip with 5 center-fire cartridges and replaced the rifle in the rack.

*****

I had passed the house once or twice a month, but until our boundary line walk I never knew who lived in it. The Squire I had seen much less often - an 80-year-old pushing the air aside aggressively as he walked to the Post Office in East Thetford. Heavy glasses, tweed coat, flannels and hat set him apart from the other infrequent pedestrians on the country road. His gait had a slight swagger, perhaps that of a person who did not want the world to forget that he had been an officer in World War I. He carried a cane, not because he was infirm, but so that he could plant it just in front of his stride. Then while passing it he maneuvered the head of the cane in a counterclockwise circle away from the body in preparation for its next planting. This exaggerated circle was the initiation of the flourish, which included the cane tip-lift and the replanting with the next stride. I later would learn that the cane offered discouragement to the few dogs that dared to follow his heels, and that it was used with vehemence to disrupt cobwebs in doorways.

From scuttlebutt furnished by some of the elders in the community I learned a bit more about him. He had arrived in Lyme in the aftermath of the great flood of 1936. The roads, bridges, and low-lying houses all needed repair, none more than the covered bridge from Lyme to East Thetford. It had been swept away, the largest section lodging in the mud where the river widened half a mile downstream. An engineer, the Squire directed the construction of the new bridge. Concrete abutments and a new center pier were constructed to support the large steel trusses that now carry the road across to East Thetford. While the depression-era laborers excavated muck at the bottom of the cofferdams, the Squire strode the job-site with a roll of blueprints and, from time to time, cast his eyes to a house up the river. This large brick house, in the classic Federal style, presided over a meadow that ran down to the river. I suspect he was thinking that it would make a beautiful residence for a man of his stature.

The house had fallen onto hard times and needed repair. The wooden fans over the doors and windows had lost their paint and hung gray and with slats missing. The shutters were, likewise, needing repair. While the Squire continued with work on the bridge, the hurricane of 1938 arrived. The storm blew the roof off the horse barn and during the next months the Squire saw its interior exposed to the rain and snow. The pastures were thin and the few cows on the place appeared gaunt. The big barn needed a new roof.

The Squire had learned that the Steele family had followed the original owner of the place, Colonel Ebenezer Green. The only survivor, Lena Steele, was the widow of Carlos Steele the eldest of three brothers who had comprised the last generation of Steeles. After Carlos died Lena had continued to live in the house with Carlos' brother, Clarence, otherwise known as "Bud".

For a time, the three brothers had been recognized as strong and promising young men. A measure of how strong they were is offered by the stone in front of the house at the west end of the Lyme Common, the former home of Dr. Hamilton, Lyme's early doctor. That stone bearing a plaque honoring the physician, used to be in the yard of the Steele house and each of the three sons could pick it up. When it came time to honor the doctor, their mother offered it to the town for the monument. No one has picked it up since.

Despite the exercise provided by lifting the stone, illness - mental depression and tuberculosis - afflicted the family. Bud Steele, the last surviving male had a hunchback, a condition possibly caused by tuberculosis. I can imagine Bud living every day with his deformity and the resulting pain. The tough life on the farm after the death of his brothers would have incubated mental depression that others in the family had manifested. Lonely and never ending work would have deprived him of other opportunities. I also see him trying do each day what he had done the year before, but finding that it was harder. I imagine him out in the horse barn as he stretched his twisted and shortened back as high as he could while he lifted the work harness down from its wooden pegs. He then hitched his team and set to work. Plowing 30 acres, mowing 10 acres of meadow, and repairing the fencing of the hillside pastures, was the real work. In addition, he milked 10 cows by hand twice a day and hauled the milk to the creamery near the center of Lyme where the cream was separated. Butter made from that cream was shipped to Boston by train. Leaving the creamery he would have returned with the skimmed milk to feed the calves and pigs. It was a treadmill of hardship that he walked. The neighboring Jenks family from further north on River Road, remembers the Steele place as a somber house as they rode by on their horses.

In 1930, Lena Steele's uncle, Bertie Burgess, delivered a cocker spaniel to the farm, a little dog named "Flash". It had belonged to his grandchild, Donald, and had bitten the young boy. Bertie decided it would be better if his niece could keep it in the country. Suddenly, there was tragedy. Flash lost all four legs in a mowing machine accident and had to be put down. The name of the person operating that machine is lost to history but I suspect it may have been Bud Steele. We know that in 1931 Bud Steele hanged himself in one of the barns. It is possible that the injury to the little dog may have been the capstone for Bud's long-building depression.

In 1939, the Squire, the only person around with any money, purchased the house and the land from Lena Steele - all 235 acres. He made extensive repairs and renovations, settled in, married a

11

music professor (a woman who used "Clef" as her first name) at a nearby college, and became a memorable figure on the local scene. The features of the house and land assumed a profile dictated largely by this peculiar person, and his wife.

Now, 30 years after the Squire had purchased it, it was up for sale. There was no sign in front - that would not have been in character with the Squire - but he had called Bob MacDonnell, a broker. For years the Squire had prohibited travel over his land. It was rumored that he feared that a virus carried by horses caused cancer and that horses would contaminate his water supply. Mike and I were now taking advantage of his absence at his college reunion for our reconnaissance. It was the only way to get a good look at the woods and fields that made up the property.

*****

We had arrived in Lyme six years earlier, purchased a 1860's house in the village and had spent the intervening years repairing and remodeling it. When we purchased it, daylight shone between the chimney bricks above roof level and smoke came from the hot air registers when the furnace ran. There was no clear demarcation between the outdoors and the inside of the house – just a series of imperfect walls. The claw foot bathtub appeared nearly new because there had never been enough water to take a decent bath. The sewer disappeared into a system of doubtful competence, and the place had been wired with cable known to cause electrical fires. All these problems had been corrected and we were comfortably settled. I dreaded the idea of starting over with another home, this one even older. On the other hand, we had purchased an additional acre of land from a neighbor, had installed three sheep and a goat to eat down the brush. We fancied ourselves as "back to the earth" and were feeling the need for more land.

12

Jean and Mike's wife, also named "Jean", had taken on the job of boarding two horses from a girl's summer camp during the winter in return for the riding they would get during the fall and spring. They were quartered on Mike and Jean's farm that adjoined the Squire's. We had looked at two other places with farmland, and had lost each one because of our low bids. Real estate values were rising and the Squire's land was much better than what we had been willing to bid on before.

Jean and I were sitting at our kitchen table under the hanging copper-plated oil lamp reproduction from Sears Roebuck. "Bob says that the Squire wants $99,000 for it. He says that we should demand a survey, and that we might get it for $95,000 if we agree to pay for the survey ourselves." The children were asleep upstairs unaware of the reasons for our distraction during the last few hours.

"But you haven't even seen the inside of the house." Jean spent a lot of time keeping me from getting too far ahead of myself.

"Sure, but the land is fine, the setting for the house is great. If the inside of the house is okay then maybe we should do it."

Jean relented, "O.K. Call Bob and we'll look at it."

I called Bob, "Jean and I would like to have a look at the Squire's house."

"It would be better if you came alone and looked at it first. Leave Jean at home. The Squire is a difficult guy." So I went alone to meet Bob and the Squire.

I pulled my International Travelall into the parking area north of an asparagus bed, alongside a large butternut tree, and walked to the kitchen door. Summer sunlight was streaming into the kitchen from the south windows, but inside it was cool, a bit clammy, and there was the smell of a poorly adjusted oil burner in the air.

The Squire was sitting at a table in the center of the kitchen. Overhead hung an oil-lantern-style light fixture from Sears Roebuck like the one over our kitchen table, his brass-plated instead of our copper. The tavern table, however, was a real antique – a single pine plank over 30 inches wide, scarred by 200 years of use.

"Good morning, Ross," said Bob, rising from the table to greet me. "Let me introduce the Squire, the owner of this lovely house."

The squire looked up at me from a body enveloped in tweed. "Sit down!" he ordered without getting up. "So you want to buy this house."

It was not a question. "Well, I have always admired it and would…." He didn't let me finish.

"Built by Colonel Green. Came up from Connecticut. Fought in the Revolution. Decided that neither New Hampshire nor New York was paying any attention to the Connecticut River Valley. Organized a new state, Dresden, right in this room." He gestured toward the fireplace as if the brick, now painted white, could confirm what had happened a couple of hundred years ago.

The fireplace was nearby. A wooden door to a bread oven was on the left. Leaning against the brick was an ancient peel, its sharp edge now rusted and twisted so that bread loaves could no longer be slid from its surface. A small black cap-and-ball pistol hung from a square cut nail driven into the mortar. The black against the white-painted brick was dramatic. Some cast iron pots sat on the floor in front of it. Maple flooring ran right up to the fireplace. There was no hearth.

He spotted me looking at the pistol. "Dick Jenks found that stuffed between two stones in the wall next to the driveway. I had him rebuilding the wall. Good man, that Jenks."

"Good man, indeed", I thought. "Instead of pocketing the pistol, he gave it to the Squire." Dick had worked for us and I knew him as a friend and an honest man. I went over to the fireplace for a better look. The hammer was immovable, rusted in the open position. The hole under the cap was also rusted shut. I wondered whether the pistol still had powder and a ball in it.

"Well, Squire, let's show Ross the house." Bob spoke in a manner suggesting that he wished to take control of the morning, but had doubts that this would happen. The Squire pushed on the table to assist as he rose, bandy legged, and grasped his walking stick. We set out – not first into the living or dining room, but into the mud-room lean-to behind the ell. It was clear that the Squire wished me to understand that he was an accomplished artist before the proceedings went any further.

We found his easel set up in the middle of the room, his paint box handy, a wobbly chair nearby. Supported by the floor and leaning up against three of the mudroom walls were about twenty canvases covered with palate-knife applied rough-textured paint in mind- jarring colors. I was appalled at the wastage of otherwise useful paint. The glaring colors on the canvases were in contrast to the brown beaverboard that covered the walls.

The light came from windows on the north and east. They were "12 over 12" windows – small panes that contained antique glass – swirls and bubbles – brought from England or from Braintree Massachusetts, the first glassworks in this country. I believe that the sash had come from an older building and had found its way into the more modern mudroom. As we walked about the room the joists supporting the floor bent easily. The floor bent with each step. The bellows-like coming and going of the floor, and the resulting change in air pressure, caused the door to the garage to flop back and forth and rattle its loose latch with every new step.

A creamed inscription, "Clef and Squire" overlay faded red and cream stripes crudely painted on the door. I imagined that it had been done for the time years earlier when the old groom brought his younger bride through that door. "Where is that bride now?" I thought, "Not here this morning."

We retraced our steps into the kitchen. In the center was the tavern table, and near it was the electric stove. To the north were entrances to three tiny rooms. In the center room the sink was against the north wall. A window over it offered a view to the barnyard. The Squire had converted the double-hung window that had been in the room into this "picture" window by having it removed and reinstalled on its side. The second of these three rooms was the pantry, and the refrigerator was in the third. In addition to the doors to the three cubicles there was the door to the mud room on the east, the door to the driveway on the south, and two doors on the west, one to the dining room and the other to the back hall. Seven doorways, exactly the same number of kitchen doorways in the house we were currently living in! Was this progress?

The dining room was a delight. Bright sunlight entered through the south windows. Deep blue wallpaper with a white flowered pattern covered the walls above the hand-planed pine wainscot. I imagined Jean's mounted photographs resting on top of the chair rail, a fire crackling in the fireplace and the family gathered around the table.

From the dining room we entered the study: shoulder height bookshelves on the north wall, another fireplace and door to the yard on the south. Over the wainscot the west windows looked down across the field to the Connecticut River. We moved on to the living room. By this time I had noted that the windows of rooms on the ground floor were recessed into the twelve-inch thick walls. Wide beveled window frames angled outwards to the surface of the plastered walls, giving a comfortable open feeling to all of these rooms. I knew that the battleship gray floor enamel, many layers thick, would prove formidable to remove, but I could see how attractive these rooms would be if given a bit of care. Even before seeing the rest of the house I felt that we should buy it.

Upstairs there were four bedrooms, one in each corner of the main house. The outside walls were one brick thinner here, eight inches instead of twelve, and the beveled window frames narrower.

A door at the top of the back stairway opened to a loft over the kitchen. Although panels of beaverboard spanned the distance between the roof trusses, the pegged mortise and tenon joints were visible. The hand-formed brick with thin lines of lime mortar that comprised the knee wall of the ell interacted with the ancient glass in the windows to provide an authenticity that was powerful. At this point I realized that I was now in the original building built by Ebenezer Green. The materials and design used for construction of the kitchen and this loft was a generation or two earlier than the rest of the house. I could now imagine the time 200 years earlier when the entire family and perhaps travelers on the way north, slept in this room. Then the Federal house was built on to the end of this building. By 1800 the riches resulting from the wool trade had made such an addition possible.

As we toured the barns the Squire led the way with his cane, tearing the cobwebs asunder and sending spiders racing for cover. In the horse barn, tie stalls, each with a small high window to the west, had provided quarters for the four draft horses. A chute from each stall ran up alongside the window and into the hayloft above. Horses' teeth had cribbed the chutes into a fragile lace-like pattern. A box stall in the northwest corner was now filled with all kinds of hand tools – rakes, shovels, scythes, picks, hammers, and the like. The top of a small schoolmaster's desk removed from its base had been spiked to one wall. It was covered with the odds and ends accumulated from multiple repairs on antiquated farm equipment. In a corner were two watering cans and a pail filled with what looked and smelled like motor oil.

"Used motor oil", said the Squire. "I mop it on the boards with a broom or brush. Keeps the wood from rotting."

The seventy-foot long hay barn had used the timber frames from two earlier barns that were brought-end-to end and finished as a single barn. Built into the slope so that the second floor could be reached from ground level on the east side, the center was comprised of two bays that reached from the lower level to the rafters.

17

On each side of these bays an entrance allowed hay wagons to be pulled in so that hay could be unloaded into the adjacent bays.

The Squire's 1940 Ford 9N tractor sat on the wooden floor at the south end of the hay barn. The heavy highway-type mower attached under its belly was angled jauntily upwards over the edge of the southernmost bay. The Squire's cane swished over it, tearing cobwebs away, his cane-tip perilously close to the unblemished gray enamel. The tires were new, purchased after the war when rubber became available again. The Squire looked at it as if he were viewing a lovely woman. "I would be willing to sell the tractor at a good price to the person buying the farm," he said.

Cut into the east side of the barn was a floor with stanchions and a feed trough at the end adjacent to the bay where hay would be stored. To the north was a floor where cows had been kept. Now it supported an ancient silage chopper as well as other old equipment. We circled around to the west and entered the barn at ground level. Behind the bays, the east wall was rough whitewashed stone. From here I could see that the floor under the stanchions had been rebuilt with new timbers. In the darkness and chill from the stone wall the Squire pointed to a single heavy iron hoop leaning against the stones. "This iron tire was on the wheel of a wagon from Connecticut. Colonel Green came up the river from there and this is all that is left of the wagon."

In one place a piece of the iron had been cut out, leaving a 2-inch gap in the otherwise perfect circle. "What happened there?" I asked pointing at the defect.

"Hired man needed a piece of iron two inches long," the Squire replied with a pained look. "That's enough for now. Bob is going to move a bed for me."

"Move a bed? Ross hasn't seen the cow barn or the corn crib," said Bob, obviously not having heard about bed moving before this moment.

"He can come back some other time. Now you're going to move that bed for me."

*****

Upon my return home I thought I should emphasize the good points.

"You are going to love the dining room, Jean. It has a chair-rail around the walls that will hold your photographs, and there is a closet in the downstairs bathroom that will be a perfect place for your enlarger."

"How is the kitchen?"

"It has seven doors, just like this one, but the bread oven will be perfect for baking loaves." Our three children were growing fast and we were going through a loaf of homemade bread a day. "It'll hold at least eight loaves at a time, and the Squire will sell the portable dishwasher that rolls up to the sink."

"Rolls up to the sink?"

I then had to explain that the sink, the stove, and the refrigerator were each in a different room. "You can wash your prints in the bathtub. It will be close to the enlarger."

Jean gave me a hard look. "I'm glad you are in medicine and not in sales."

Later in the evening, when the children were in bed, we continued our discussion, this time working over the financial assumptions. We had three children. I assumed that all of them would be headed to college. Jeanie was ten years old, Ross eight, and Elizabeth four. We had to consider what college would cost. I had joined the faculty at Dartmouth Medical School one year after the decision was made to cap the scholarships for faculty children that were provided as a fringe benefit at Dartmouth. Since that time tuitions around the country had risen rapidly.

The anxieties that lay beneath the surface had to do with taking on debt. This unease was encumbered with the remnants of Depression children's guilt. It arose from the future ownership of a gracious house in a lovely setting. Were we reaching too far? Neither of us believed in divine retribution, but we both understood that the downfall of the arrogant makes for great village gossip.

Mike and Jean's land bordered the Squire's and they wished to protect the view from their kitchen table. They wanted an old pasture with a few wild apple trees providing shade and fruit for deer that lay in the northeast corner of the property. There was access to this parcel from the town road and we figured that we could sell it to the Smiths for enough to make our purchase feasible.

I went to the bank, and to the Treasurer's Office at Dartmouth to inquire about mortgages. My temerity about the economic future was not shared by either when they learned that we would be prepared to make a substantial down payment on the Squire's place. The college offered the best terms, six percent, for 30 years.

Despite this reassurance, for a number of nights I had a recurrent dream in which I entered a large house through an open basement door. I wandered about a pit in the basement that held a huge furnace and then climbed a complex stairway that led ever more steeply to a room in the attic. I worried all the while that I didn't own the house or belong there.

Chapter 3

Inspection

B ob, how was the bed moving?" Although a real estate broker is the agent of the seller, I hoped that Bob might be on the side of the purchaser by now.

"My back will never be the same again! It was the heaviest bed in Lyme, and it had to come down those narrow back stairs. "

"I think we are going to do it, but Jean hasn't been in the house. Tell the Squire that she is coming over."

Jean went over while I stayed with the children. I was delighted with her enthusiasm upon her return.

"Okay, we will make an offer, once I've had a look at the last barn and the corn crib."

I parked the Travelall next to the asparagus bed again and headed for the kitchen door. No one was at home - an opportunity to look things over without supervision by the Squire. I had a quick tour of the remaining barn, a long structure forming the north side of the barnyard enclosure. Two bays were open to the south, and a small room with stanchions on the west was deep in dried cow manure. Climbing a short steep stairs in this room put me in the hay-mow overhead. It was floored with pine boards an inch thick laid over widely spaced joists. I stepped carefully, avoiding a plunge through them onto the ground, which I could easily see through large cracks and missing boards.

Descending, I crossed the open bays and entered the small room on the east end of the barn. It was whitewashed and airy. Light streamed in through the south-facing windows. Under them was a row of stanchions. A door in the east wall led from this room into a small yard. The fencing for the yard lay on the ground, the posts rotted off. Pollen from the heads of a tall stand of timothy covered me from the waist down as I crossed this yard. The grass was so tall that I did not notice the flattened area until I reached the center of the yard. Here, where an animal had been bedded down below the tops of the grass and hidden from sight of the house only a couple of hundred feet away, I found a pile of partly dried bear scat.

Now alert, I cast my eyes toward the corncrib fifty feet away. There was no sign of the bear and the track from the bedded area led away to the east.

The corn crib stood tall and narrow. Its base was supported on posts, each sheathed in galvanized tin roofing. The tin had been continued up to cover the entire first story as defense against rodents. I went to the metal covered door and removed the wrought iron pin from a beautifully handcrafted iron hasp made by some long-forgotten New England blacksmith. The wide door swung outward on long strap hinges probably made by the same blacksmith. Just inside the door a stairway led to the second floor. Next to the stairs a chute descended from the floor above. A cobweb-encrusted barrel had been placed to catch what came out of the chute.

The post and beam frame was hemlock. The main part of the frame was solid, with tight joints and capable bracing. The joists supporting the second floor must have been chosen by a different crew. They were of odd sizes and shapes. The face of the eight-inch post to the right of the door displayed the holes drilled by a shot gun fired close. Rats, I supposed.

22

On the floor were about fifty rough sawed hemlock two-by-fours along with some two-by-sixes and random pine and hemlock boards, mates to those that had been used to install the beaverboard in the room over the kitchen. Up above, a corn sheller was positioned over the chute to the first floor. The boards forming the walls had elongated cutouts for ventilation and the entire second floor had been wrapped with wire mesh.

A large window on the second floor looked out to the west, and the sun beamed in through the two south windows on the first floor. If cleaned up and put on a decent foundation, this building would make a cozy living area with sleeping loft above. My anxieties about finance were suddenly relieved. If we ran into problems, we could subdivide and live in the refurbished corncrib as we rode out the difficulties. I stepped out of the corncrib having made the decision to buy the place, even if it meant paying the Squire's full price.

To the north of the corncrib was a gully holding the small brook Mike and I had crossed when we inspected the farm from the overlook to the east. As I left the corncrib and entered the shade cast by a heavy growth of trees along the brook I could hear falling water. I went out on a small ledge to watch the water plunge about five or six feet, strike a ledge outcropping and splatter another dozen feet to the mossy rocks below.

*****

"Bob, we will make an offer on the Squire's place."

"Before you do, let me tell you that he has set a price on everything in the barns, including the tractor. It might help you with the decision of what to offer. He wants $800."

"That is a very good price. We will offer him $95,000 plus $800 for the barn contents and he does not have to obtain a survey."

"He also will sell you his truck for $300."

"Fine, I'll take it."

23

"I'll recommend that he accept the offer and get a purchase agreement done up."

"Bob, I would like to stroke his ego a bit. Tell him that part of the deal is that we want one of his paintings to go with it."

"I'll have the purchase agreement for you to sign tomorrow night after work."

"Tomorrow night, Bob, I will be the host at a shindig I am giving for the 200 people attending the Leukocyte Culture Conference that I'm chairing. The conference begins tomorrow morning. My fellows, who were supposed to help with the inpatient service, have both had to leave early, and I don't know which end is up. I'm going to have to see my patients beginning at 5:30 AM, finish at 7:00 AM, make introductions for people from all around the world, and then stay awake even when the room is dark for slides."

"Don't worry, I'll track you down. It won't take a minute."

With considerable effort, I managed to stay awake in the darkened auditorium while the conference participants described the studies that opened up what was then a new field of study, cellular immunology. We adjourned in the late afternoon to an outdoors recreation area nearby. Jean met me there while the international group's bonding was aided by volleyball, swimming, hamburgers and beer. As the sun went down, the keg went dry. I took it off to the store to get another. I returned to find Jean and Bob looking for me.

"Sign here."

I signed. Jean had already signed.

The conference continued for another one and a half days, and I scarcely thought about the Squire or his farm until the last speaker had finished and I said farewell to the conferees.

*****

Jean and I sat at the kitchen table with the three children. "You know I promised you that we would buy a television set so that you could watch the moon landing?" The children knew that I intended to make good on this promise because we had just finished converting the woodshed next to the kitchen into paneled den with a cabinet to contain the set. "Well, Mom and I have decided to buy a farm instead. We will go over to Bunny's and watch the moon landing on her TV." I expected the worst. Instead, the reaction was joyous.

"Can we get a pony?" "When can we see the farm?" "Does it have a barn?" "When do we move?"

"Yes, we will get a pony, if you agree to take care of it. That is why we will not get a TV. Instead of watching TV you will need to help your mother and me do the work. We will go over to see it next weekend. It has some big barns. We will move in September." There were many more questions, of course, about rooms, access to the river, the brook, gardens, and the forest. Part of this discussion dealt with other things that would have to be put aside if we moved to the farm. If they had ponies or horses, for instance, they would need to feed and take care of them.

"We are also buying a pickup truck and Jeanie and Ross will learn to drive it on the farm."

Elizabeth looked hurt. "And Elizabeth, too, if the truck lasts until she is ready to drive it."

The next weekend we were met by the Squire who admitted Jean and Jeanie to the house (after ascertaining that Jeanie had clean shoes). I was told to wait outdoors with young Ross and Elizabeth. He thought Elizabeth too young and Ross too rough to be admitted. While we waited, Ross danced along the top of the stone wall that Dick Jenks had built, loosening a few of the rocks but uncovering no more pistols.

Elizabeth and I wandered along the driveway that circled the rear of the house. I heard the kitchen door open and the Squire's gruff voice directed at Ross.

"Get out of that tree, you don't own it yet!" Ross sheepishly let himself down from the low spreading limbs of the big maple adjacent to the stone wall.

"It's O.K. Ross, I know you were not doing any harm to the tree or anything else. He is just that way." I said as Ross joined Elizabeth and me. We continued our amble around the drive and inspected the horse barn until Jean and Jeanie emerged.

"Dad, Mrs. Sessler is going to give us pictures of Beethoven and Mozart. They have a nice piano."

"That's wonderful." I replied, thinking, "Let's get out of here, while we are still alive."

*****

Later we described our outing to Bob who was amused. "Bob, what about the painting?"

"Ah! The Squire said to come over and make your pick. Give him a call."

It was on this expedition that Jean and I first toured the house together. First we looked at various items that the Squire and his wife, Clef, had decided not to move to Florida. The large spinning wheel in the kitchen, its spindle missing, was too big to move. We bought it, the pistol that Dick Jenks had found, as well as antique tables, an ancient six-board pine blanket chest, an antique maple child's bed and other furnishings that came to a total of $571. "Now let us look at the paintings," the Squire beamed. "This one you can have for $800," he pointed at a vivid abstract that hung on a dark wall.

Jean and I looked at each other incredulously. This is not how we had intended to stroke his ego. Had Bob miscommunicated and suggested that we would pay extra for one of these outrageous canvases? Not knowing what had actually been said, but understanding each other's wishes and goals, we toured the house looking at one painting after another. If I said I liked it, Jean always found something wrong, and vice versa. Sometimes we both found the painting inappropriate. We never both agreed that we liked a painting, although as we viewed one after another the prices gradually came down. Finally, the tour was complete, and we had not identified a single painting we wished to purchase.

"Well, I guess we are going to have to go away empty handed," I said.

"What about the painting over the fireplace in the dining room?" Jean asked. This painting was clearly from the last century and was of a stream crossing with the sun setting over it. Not painted by the artist in front of us it was displayed in a heavy gilt frame and buried beneath layers of darkening varnish.

The Squire sighed as if dealing with imbeciles. "That painting is not mine, but I will let it go with the house."

"Oh!" Jean responded innocently. "I thought it was yours, but we will gladly accept it."

A week or two later, Bob called us. "You know that $800 will be added to the purchase price for the items in the barn."

"That's right."

"Well, the Squire just had me lift down the cutter that was up on the rafters of the garage. The person who bought it, helped."

The cutter had been up there for years, and was covered with dust, its seat cushion in need of repair, but otherwise all set to hitch to a horse on a snowy day.

"That cutter is ours!"

"I explained that it was yours, but the Squire said that he sold you the things in the barns and that the cutter was in the garage."

"I assumed...."

"So did I, but I think he has got you. I suggest that you get over there and make a list of what is in the barns."

While there, I got permission from the Squire to mow the hay and put it in the barn so that we would have it for a horse. Clyde

Grant, who cut hay on many places in town, was too busy but agreed to rent his equipment to Frank Cutting who could do the job. The grass was dry and yellow, far past its prime, but it was waiting for us.

Frank mowed with care using Clyde's Farmall H tractor and the dust curled up from the brittle grass. He returned two days later to rake, producing long windrows of hay, which coiled along the length of the field behind the house. When I made the dash to the farm during my brief lunch hour, I found Frank at the wheel of the Farmall pulling Clyde's New Holland baler. With each punch of the ram a cloud of dust emerged from the machine and drifted toward the brick house. As each bale dropped from the chute more dust was cast into the air. Lounging near the driveway, sipping cold beers from a cooler, were three young men that Frank had lined up to pick up bales. I was here to pay them.

One of these fellows was fresh from service in the jungles of Vietnam. His eyes, face, movements, were different. I gave each of the others a ten-dollar bill for their day's work. The dark-faced fellow from Vietnam got a twenty, since I was out of tens, and I told him that I was paying for his time tomorrow, as well.

Early the next morning I got a call from Frank. "Wanted to let you know that the guy you gave the twenty to yesterday is dead. Smacked a tree last night when he was blind drunk. The Squire won't let us finish. Say's it is too dusty. I think the real reason is that he didn't like the guys drinking beer, especially the man who's now dead."

Those long windrows rotted on the ground while I wondered if I had found another ten dollar bill instead of the twenty, the Vietnam veteran might have lived.

*****

We were now approaching the date for the closing on the house and I went over to purchase the Squire's truck. I sat down at the kitchen table across from him. Clef was working at the kitchen sink. The squire turned toward Clef, "Go out and pick some flowers, we have business to do." Clef quietly left the room and the Squire said, "Now, that was $300 for the truck, correct?"

"That is correct."

"I just checked the gas gauge. It is a 15 gallon tank and it is half full. That is seven and a half gallons at 28.9 cents a gallon or about two dollars. Write a check for $302."

I pulled out the checkbook and wrote the check. I left the blue 1950 Chevrolet pick-up with its curved quarter round rear windows in the garage and hurried home in the Travelall. "Jean, listen to this. The Squire charged me two dollars for the gas in the tank!"

Jean gave me a hard look. "And you paid it? You dummy! You should have told him that you bought the truck, not the gas. You should have told him to get down underneath it and drain the gas out!"

Finally, September 16, 1969, the day for the closing arrived. I came home from work in mid-morning to pick up Jean and to take her to the Treasurer's Office at Dartmouth College where the closing would take place. As we drove toward Hanover, I spotted a car speeding in the other direction. As it shot by, I caught a glimpse of Bob at the wheel. "Hey! That was Bob headed north. I wonder what is going on. He should be in the Treasurer's office."

We were taken to the Treasurer's conference room with its fine collection of solid furniture in a high ceilinged room that gave the impression of stability, wealth, and good judgment. The Squire was already sitting at the table and appeared to be in command of the situation. The other principals were there also. The Assistant Treasurer gave me wink that the Squire did not see. "Ross and Jean, I just had a call from your real estate agent, Bob. He will be just a few minutes late and suggests that we wait for him."

We sat for nearly 40 minutes making small talk, occasionally interrupted by the Squire demanding to know why Bob was so late.

At last, Bob entered the room with Clef on his arm. With a broad smile he gave a warm greeting to the Squire. "Squire, I'm sure you know that this deed will not be any good unless Clef also signs it, so I went to pick her up."

The squire, dumbfounded, looked at Jean and me, then at the others in the room with signs of increasing resignation. He didn't glance at Clef. We signed, passed checks, and we now owned the place. As we signed we also completed documents transferring 67 acres to Mike and Jean. We had done it! The celebration came later, however. Jean left me at my office at the hospital, drove home and returned at the end of the workday to pick me up.

Chapter 4

Moving In

The first reds and yellows of fall were showing up in the maple leaves as young Ross and I bounced along the road in the newly purchased pickup. He sat propped forward on the slick leatherette seat, proud to be riding in a truck. With exaggerated hustle he had helped me slide cartons of books along the smooth steel plate that the Squire had installed to cover the rotted planks in the bottom of the pickup box. "What do you think, Ross?"

"I think it is a pretty good truck."

We rounded the sharp corner on the East Thetford road, the narrow tires complaining under the load, and headed down the hill. The Chevy "Thriftmaster" engine was glad for the rest. I had noted an extensive oil stain on the floor of the garage where the truck had been parked. The drip came from the worn out rear main bearing seal. One of my recent purchases had been a ten-quart can of oil. "You are going to have to ride the school bus now, instead of walking to school."

"That is O.K. It will be worth it." The young boy spoke with great seriousness as if he were twenty-five instead of eight. As he peered ahead, eyes actively scanning the road for hazards, he was acting as a vigilant co-pilot with a full share in the outcome of this endeavor. Moments like these made me regret the many evenings during the last several years I had returned to the hospital for an additional couple of hours of work. I resolved that I would cease the evening work and, instead, go to work a bit earlier in the morning. Life with the children was too good to miss, and the farm would provide a justification for a change in schedule.

We arrived at the farm to find Jean, Jeanie and Elizabeth unloading clothes from the Travelall and carrying them into the house. The only closet on the ground floor was a small one under the front stairs. The size of the closets upstairs indicated that during the early 1800's people had very limited wardrobes. Many houses of this period lacked them - an entire wardrobe could be hung from a few nails or pegs in the wall. The significant wealth of those who lived in this house was indicated by the presence of these closets. We considered it lucky when our own limited wardrobe was accommodated.

Later, the moving van arrived with the heavy items, and backed up to the exterior door on the south side of the library. Painted shut and stiff on its hinges, this door was opened for the first time in many years, while our beds, bureaus, tables and chairs were carried in. As this was going on, the children raced from one room to another, out to the barns, and around the house, making one discovery after another. I gathered the kids and took them through the mudroom to a narrow door in the rear wall. "I want you to agree on how this will be used," I told the children as we peered inside. They surveyed the arrangements, two full sized seats and one sized for a child, lifting the lids to see the shallow pit below and a clean-out door to the outside. The walls were a collage of faded color illustrations from the Saturday Evening Post, and Colliers.

"We could put our skis in here!"

"And we have some posters that could go on the walls!"

"We could use it, if we run out of water!"

I found a high tailgate that could be chained to the sideboards of the pickup truck box and returned to our old house with all hands to round up the sheep and their lambs.

We hooked the ewes with a cane once used by Jean's grandfather, holding them until we could get a hand under the chin and one on the butt and push them into the truck. Those that escaped the cane, replaced their stiff-legged joyful hopping by desperate acceleration. That lasted until they hit the fence and stuck with their heads through the wire mesh. The moment before they can back away from the fence and escape is the critical one in lamb catching. Our timing was perfect and the truck was soon loaded.

The next weeks were a hectic time of sorting out, establishing new patterns and learning. We walked portions of the boundary, put in new fence-posts, picked up old fences, mopped the hemlock planks from the corncrib with creosote and spiked them in place around paddocks. We ordered rolls of new sheep fencing and a freeze proof water-bowl for livestock, established that the grain bin in the large barn was indeed rodent proof, and talked about ponies, horses and cattle. Our sheep and lambs exhausted the coarse timothy in the barnyard, and we got the field to the north of the barns fenced just as they finished. That too, was not going to last long, so we began the task of fencing the best of the hillside pastures.

Up to this time we had pointed to various parts of the farm and had given general descriptions of places when we conversed. It was now time to name things. We sat around the kitchen table one evening and proceeded to name as many things and places as we could. I had made a crude sketch of the farm boundary, and had placed the major features upon it. As we reached a decision about a name the children wrote it on the sketch. This annotated drawing became our atlas. Soon we and the neighbors used these terms as we talked about the property.

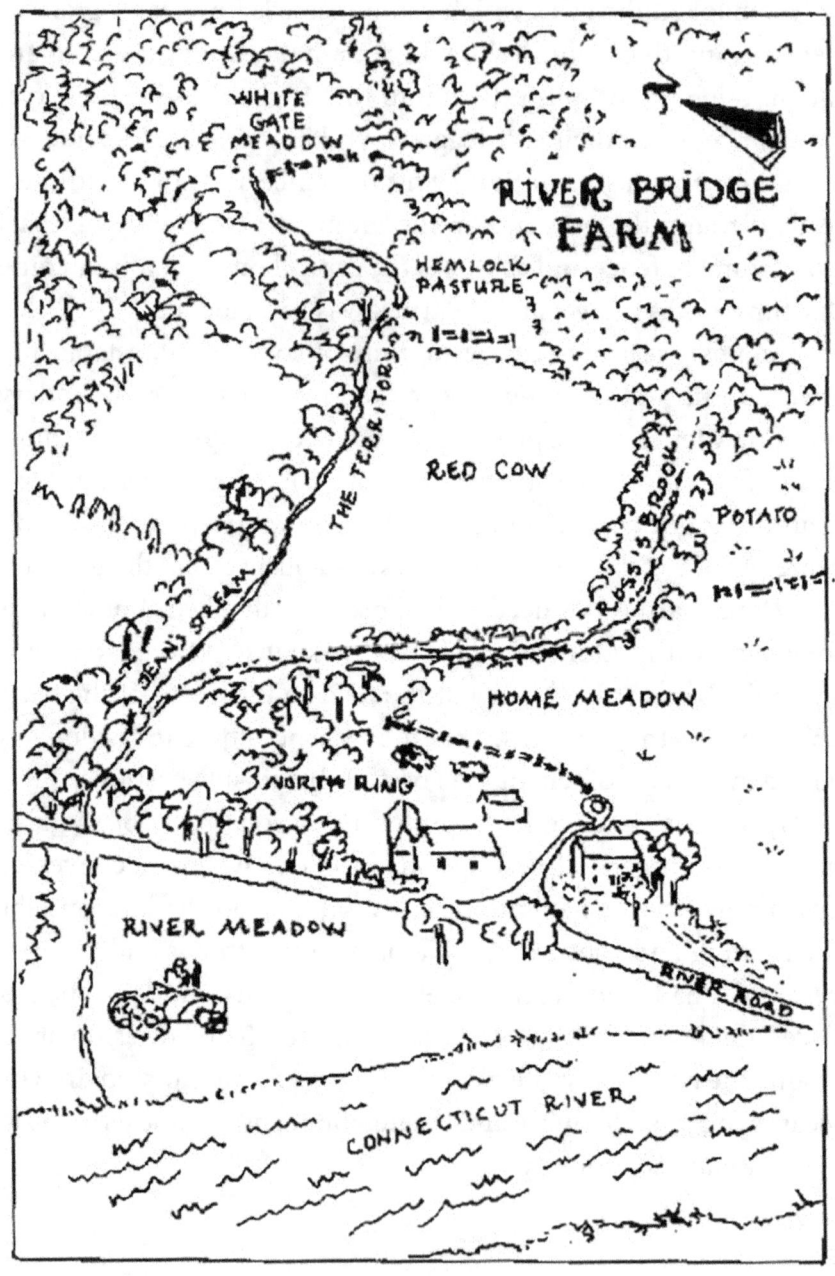

As a child I loved finding maps in the books I read. The one we made of Riverbridge Farm was lost so I asked Kate Emlen to imagine one and draw a replacement

The large field between the house and the river became the River Meadow; the one surrounding the house, the Home Meadow. The brook that emerged from the woods, splitting the hillside meadows became Ross's Brook, named after the younger Ross, and its major tributary, Jean's Stream. The field to the south of the brook was named the Potato Meadow, since years ago potatoes had been grown there for sale to those in Lyme and surrounding communities. That to the north of Ross's Brook was named the Red Cow Meadow after the Hereford cattle that we planned to put there. Further up the hill from the Red Cow was a small pasture broken up by steep rock ledge that sported clumps of ground hemlock. This was named the Hemlock. It was from the top of the Hemlock that Mike and I had first peered down to the Squire's house.

It was a steep haul up from the barns through the Red Cow Meadow to the top of the Hemlock and the workhorse path used by the Steele farmers took a tortuous route in order to lessen the grade. This placed the old gate from the Hemlock into the next pasture down near Ross's brook instead of on the mounded top of the pasture. In the cool shade of large pines and hemlocks a white-painted gate, long abandoned, was slumping into the ground as its bottom rotted away. This next pasture we named The White Gate. It runs almost level to our boundary with Mike and Jean's farm. The woods road that Mike and I had encountered on that first exploration runs from the White Gate across Ross's Brook through a thick stand of young pine to an old log landing. Here, a log crib had been constructed so that logs could be rolled off onto trucks during a logging operation in the 1940s. At this point, which we called the "Logging Crib" the road split, one route headed for Mike and Jean's land and the other, turning sharply south, coursed through a grove of ancient maples and up a steep hill. At the top young pines and oaks surrounded a rock-strewn pasture, the highest land on the place. This we named simply, the High Pasture. Beyond it, lost in the big pines, was a small slit-like pool Mike and I had passed while

35

reconnoitering during the Squire's absence. We called it the Indian Ocean, since we fancied that Indians at one time might have eaten meals on the smooth ledge overlooking its northwest shore.

We named the barns for their functions - the Horse Barn, the Hay Barn, the Cow Barn, and the Corn Crib. The flat field to the north of the barns, once fenced, became the North Ring since I would harrow a circle of soft earth for horseback riding. The falls in Ross's Brook behind the Corn Crib were named Hematology Falls after my medical subspecialty.

In future months we named many other features on the property. For instance, the bridge across Ross's Brook was named the George Washington Bridge. When rebuilt during the crisis of Nixon's Presidency in 1974, we named it Impeachment Bridge. When the bridge planks rotted and were replaced this was the "Imelda" recon-struction. The ravines and promontory to the north of Hematology Falls became simply, "The Territory".

The sessions at which names were applied and the use of these names thereafter had an effect that I did not anticipate. We were, by consensus, developing a family code. Although we could share it with our friends when convenient, and friends could use part of it, the full code was understood only by us. The daily use of these names reinforced the meaning of these abstractions for place. Each use of this shorthand annealed the familial bonds. For instance, near one path there was a dead elm tree so large that it defied easy removal. As we passed under it we kept a wary eye upwards lest falling branches hit us. It became the "Widowmaker". Now, though it is long gone, I could bury a treasure where it once stood and if, in my final instruc-tions to family, I write, "It is under the Widowmaker," the children will know where to dig. There is a place in the center of a woods road where each year several clumps of delicate Columbine bloom. This happens despite boot treads, tractor tires, and horse hooves. "The Columbine are out," we say and each of us knows that place and those flowers. A large low horizontal limb of a pasture pine was scraped smooth by the backs of many scratching cows. When Jean said, "the wind last Saturday blew down the cow scratching tree," I knew the tree and could almost feel the patina on the bottom of that limb.

Soon, the children had constructed a retreat between the large logs forming the three sides of the log crib. Excavating the soil that formed the back and bottom of the structure with care not to undermine the crib itself, they carved out a compact bunker that would hold three children. They carried some old steel roofing up through the Red Cow Pasture and then along the woods road to cover their "fort". From this secure and weather-tight redoubt they scanned the vacant intersection of woods roads - invisible observers of imaginary comings and goings. Each day the children emerged after hours spent at enlarging the excavation. Healthy teeth smiling through layers of quality topsoil greeted us upon their return. Jean found that silt remained in the bottom of the washing machine after putting their clothes through it. The insoles of their sneakers grew thick with a layer of pounded clay, a layer only marginally thicker than what was behind each ear.

The children's delight at having a place of their own, where no harm could be done, and where the range over which imagination could roam was boundless, offset the minor annoyance of a trail of fine dirt through the house.

As the family learned to generate meals in a kitchen that had the sink, stove, and refrigerator, each in a separate room, we recognized the extra burden that Jean had taken on. The main part of the kitchen had plenty of space and we had put our Montgomery Ward wood stove in front of the closed-off fireplace at the east end of the room. Its stovepipe was inserted into the brick chimney. The presence of this stove, a cheap poorly made item in the historic room was just another annoyance for her to deal with. Fortunately, we could get rid of it.

The Story of Bill Clark's Woodstove

Bill Clark was a dairy farmer. His place was north of us on River Road (The Drive-In Theater was in his back yard). Bill went off to World War II soon after graduation from Dartmouth. In July 1945 he was on board a troop-ship with soldiers who were going to land on the Japanese Home Islands. The Hiroshima Bomb, and the surrender that followed, kept him and a lots of others from that ordeal. When he got off the boat he decided that he wanted to be a farmer, not a business man.

Bill liked to golf. I don't know much about golf, but I'm pretty sure that Bill was a good golfer. Some days after lunch, Bill would let his barn full of cows listen to the local radio station and cross the river to Fairlee, Vermont. There he would play 18 holes and return to do the evening milking. One of the nearby farmers has his cows listen to opera. Why not?

When Bill's wife, Martha, became seriously ill, he needed help to get the work done. I did what I could to help. It was corn harvest time. Bill drove the chopper and I drove along-side Bill's machine in his old dump truck while it was filled with corn for the silo. The seat on Bill's tractor was broken so that it couldn't be made level side to side. I could see how hard it was for him to sit on the crooked seat. It made me think about how hard it was to pay hospital bills when your wife is sick and there is no money. I should have bought him a new seat. But, if I had, he might not have installed it on the tractor. He didn't rank repairing equipment higher than playing golf. Lots of things Bill had were broken. It seemed to happen a lot.

One day he showed up late in the afternoon at our house. He was driving Martha's car.

"Ross I need to borrow your pickup truck. Mine is being repaired and I need to deliver a calf to Enfield. I'll be back in an hour."

We had just replaced the Squire's old blue Chevy with a new truck. Dread rose along my spine. I could see the lovely truck getting

old before my eyes. I swallowed hard. "Here are the keys."

Later that night he was back. "Ross, I broke your windshield. But it is O.K. Your insurance will cover it."

Without thinking, I replied, "Bill, I have $100 deductible."

"I'll tell you what. Out in my shed there's our old woodstove. It is in parts but it is all there. It works fine. It is yours."

Jean and the children worked on the pieces with steel wool and stove polish and before long we had a new/old wood stove.

Bill's stove was put a few feet from Jean's electric stove. It had a large fire-box, warming ovens and an oven that was sometimes used by the children to warm their feet. Jean's smile bloomed as she started to use it.

The small amount of reading we did at this time was highly directed. For instance, our friend, Bunny Horton, who must have perceived the stress we were under, lent us a copy of one of Eric Sloan/s books that is based upon excerpts from the 1805 diary of a New England farm boy. Here we found that the rhythm of farm life we were just beginning to sense was fully developed in the life of that century. During the cold evenings of winter, when the feeding, milking, cream separation and other tasks had been completed, Sloan mentioned how the harness could be repaired and readied for spring plowing. The roof could be shingled during the night when the November moon was full. By that time the crops were in, the daily chores completed, the fall rains over, and the moon was suspended in crystal clear air so bright it cast shadows of those who worked on the roof. After a hearty laugh at this bit of lore, we sheepishly developed a list of dozens of things that could be accomplished on a dry, cold, moonlit night in November.

*****

One June, a number of years later, as the fields were bursting with clover blossoms and the farm animals fat and sleek after a well-fed winter we were visited by Makio Ogawa, one of my most talented medical researchers accompanied by his wife and his father. As they watched, the calves and lambs cavorted in the richness of the day. Makio's father, an English teacher from Osaka was a scholar interested in the works of Willa Cather. Because I had grown up in Nebraska this Japanese teacher made an instant connection between Willa Cather, rural life, and me. For him an encounter with a person who pursues a dual career in medicine and farming was a novel experience, something that would be most unlikely in Japan. Makio's wife, a New Hampshire woman, was less impressed, of course. On my part, I was delighted that on the occasion of their visit, we were able to walk along the farm paths at a time when the

40

fences were straight, the cows had not broken out, and the weather did not threaten any critical farm operation.

As we strolled past the various pieces of farm machinery and I pointed out the long-term goals we had for the place, Makio was impressed at the knowledge we had acquired about animal husbandry, crop production, machinery and many other things. Knowledge on these topics had not resulted from a deliberate attempt to learn, but rather from coping on a day-to-day basis with things that needed attention.

Makio was particularly impressed with the machinery. He had arrived in the U.S. never having operated a motor vehicle. As a young medical trainee he had recognized that life in his new country was going to be miserable unless he acquired a car and learned to drive. With characteristic directness, he soon purchased a car and arranged for a Korean surgical resident to teach him to drive. This training was less than comprehensive and on the occasion of one of their first outings the car was wrecked, a total loss. In fact, as I walked down the hallway between the emergency room and the X-ray department that day, I spotted a litter bearing a disturbingly pale familiarity. This generated a closer look that allowed me to recognize the frightened young man. "Makio, what happened to you?"

"Learning to drive," Makio said with a resigned expression that suggested that the pain of learning to drive was inevitable for all that took on this task. He recovered from his injury, purchased a new car and was soon back at the painful process. This car too was destroyed in an accident only a few weeks later. With the good sense to purchase a car of the same make, model and year of the second totally wrecked car, Makio then set out on the road a third time and found success.

As we strolled along the farm paths that day, Makio asked, "Ross, how did you learn about all this farming - the machinery - the animals? Did you grow up on a farm?"

41

"No, Makio. Neither Jean nor I grew up on a farm. We are both city people. We learned our farming at the school of hard knocks."

"School of hard knocks?" Although Makio's English was fine, he was still in the process of learning the idiom.

"Like you learned to drive," His wife, Mary Jane, answered.

*****

Before winter set in, we had several goals. I had promised Jean that we would get a horse, and we had decided that we would get some beef cattle. For both to be feasible we had to get ready with fences.

Our County Agricultural Agent, Dick Rutherford, came by to give us advice. He looked over the thin carpet of native grasses and weeds that had gone to seed in the Red Cow pasture. "Might support one cow if you fenced it."

"One cow!" I had visions of a slowly moving herd grazing on the hill. "Oh, I will fertilize it!"

"That will give you more of what you have there now, but it won't make a good pasture. That field needs to be plowed, limed, fertilized and reseeded. Then you will have something worth while."

It was clear that these tasks were inconsistent with my work at the hospital, medical school and the research lab. We decided to fence more of the fields than we had originally planned and to cut back on the planned cow herd. Then we would do some plowing and seeding in the spring.

From the edges of the Red Cow Pasture, I felled the many overhanging trees that had started to encroach upon the open space, and cut their trunks into posts 6 feet long and four to eight inches in diameter. We dug holes for the new fence posts with two post-hole diggers. When our efforts fell behind schedule, we hired three strong high school students who dug, carried and tamped in posts as the fence snaked around the field. Things went well until they arrived at the edge of Ross's Brook at the lower end of the Red Cow pasture. There, the clay was heavy with moisture and stuck in gumbo blobs to the diggers. We also had failed to provide a ready source of sweet carbohydrate to give the three laborers some respite. Suddenly both posthole diggers had broken handles and the tired boys were headed for home. When I found the abandoned project upon my return from the hospital I made one of the diggers whole by salvaging a remaining good handle and using it as a replacement for the broken handle on the other digger. I finished the hole digging job as the fall moon rose, and then after dinner, carried the posts, most of which were heavy and wet soft maple, to the remaining holes and tamped them in.

We later found that the maple stems were the worst possible material to use for fence posts. Although buds and new leaves soon sprouted from their sides in areas where the soil was moist, of course they never took root. Moreover, the underground portions were perfect nutrition for wood decaying fungi that destroyed the lower 2 feet in short order. The hard wood above the ground resisted the entrance of fence staples so that Jean got tennis elbow from hammering them in. Only 5 years later when the oil embargo of 1974 arrived were we repaid by that first year of fencing. As fuel prices soared, we had a hundred four foot maple posts rotted off at the ground, dry as a bone as they swung from sagging barbed wire and ready to be harvested for the wood stove.

Early in September I had ordered a freeze-proof water-bowl made by the Nelson Manufacturing Company in Iowa. This device, displaying the design simplicity that only genius confers, works without complaint, year in and year out, furnishing a reliable source of drinking water for animals. When it arrived, I removed it from its carton and placed the gleaming stainless steel and aluminum device near a recently purchased five-foot section of glazed clay sewer tile. The tile would be set on end in the ground to protect the pipe that rose to fill the bowl and the stainless steel base of the device would be fixed into the bell-end of the pipe with mortar. All that was necessary now was to dig a ditch 5 feet deep for the water pipe, below frost line in our area, from the house to the barnyard - about 150 feet.

Dick Jenks had promised to bring his backhoe for the excavation. However, there were delays. Dick always offered a good explanation: the backhoe was busy with a job a number of miles away, and a new job, just next to that one had come up. It didn't make sense to transport the backhoe all the way to Lyme and then take it back to a job that was so near at hand. When the backhoe was in Lyme it broke a hydraulic hose, and then there were emergencies - a septic tank failed, a water line broke. Meanwhile autumn's chill wind began to blow, a skim of ice formed in the shallows along the riverbank, and the ditch was still only some chalk on the ground to mark its ends.

Our continuing fence work provided a daily estimation of the frost level in the ground. On some days the cutting edges of the post-hole digger hammered down through an inch thick disk of frozen soil. Then the sun warmed things for a few days and the ground thawed making the job easier. Soon, however, the ground froze deeply enough so that even when the sun shown brightly only the top inch or two could be easily dug. Below this level the ground was as hard as cement. My calls to Dick became more and more insistent and his promises were correspondingly more emphatic but also unfulfilled. Just before Thanksgiving, he promised to come the Saturday after the holiday, but didn't arrive.

That Sunday morning, I arose early and drove to his house. Ann, his wife, had just entered the kitchen and had put the coffeepot on. "Hello, Ann. You know who I'm looking for."

"Yes. He still is in bed."

"I hope you won't mind, but I'm just going to sit down in the kitchen and wait until he gets up. Then, I'll collar him. If we don't get that ditch dug today it won't be done till May."

Half an hour later, Dick clumped down the stairs and stared at me with disbelief as I sat at his kitchen table. "Ya got me beat," he said as he put his fingers through his tousled hair and sat down for a cup of coffee. A few hours later, the ditch had been dug, the pipe laid, and the trench back-filled. We had water.

Chapter 5

Cattle

To make the farm profitable with the limited time we could devote to it we decided a beef cattle operation was the way to go. Walt Record, our neighboring dairy farmer, owned cows that frequently broke through his fences and roamed over our high pasture, leaving unmistakable evidence of their passage. I realized that if his Holstein heifers found nutrition enough there, that beef cattle would also.

Nor was there any doubt as to what breed of cattle we would get. My father spent his summers on the farm where his aunt and uncle lived, Egleton Court, near Ledbury in Herefordshire, England. This farm raised Hereford cattle and photographs of championship bulls adorned the walls of the dining room. Fine bulls had been exported to establish the breed overseas, particularly in the U.S. and Argentina.

When my father received his degree in Medicine at the University of Chicago in 1932 it was fortuitous that he was appointed to a faculty position at the University of Nebraska Medical School. Here his salary was derived from a state budget that depended to a large extent upon corn and Hereford cattle. (Black Angus cattle came later.) The white-faced cattle, animals whose genes had originally been selected and congregated by cattle breeders at Egleton Court and its neighboring farms, grazed on the rich grass of the Nebraska Sand Hills, were fattened in the feed-lots of the Platte valley, and were shipped to the acres of pens of the Omaha stockyards.

So Herefords it was. For several weeks that fall, I crossed the river on Monday nights to attend the East Thetford Commission Sales auction in order to get a feel for what animals were worth and to look for suitable livestock. Livestock auctions perform the same role for farmers as the New York Stock Exchange does for investors. Once a week in Thetford, instead of daily on the stock exchange, willing sellers and willing buyers were brought together to determine the value of what was offered.

The auction company, owned by Carleton W. Gray and his family, was based in a large barn and attached sheds, surrounded by numerous pens. Carleton's wife, sons Larry and Herb, and Herb's wife along with numerous part-timers comprised the help. A sawdust covered arena was in a large lean-to to the south, warmed by a large suspended propane heater, and surrounded with several tiers of rough pine bleachers. Access was provided by one of three doors that also were used by the animals. A well-worn coat, rubber boots, and a relaxed attitude were required in order to feel comfortable in the place, and unless one was comfortable, the considerable education in farm economics, animal husbandry, and human behavior offered by the establishment would go wanting.

By the third visit to the auction, I could sense relaxation as I stepped through the opening of the large sliding door on the east end of the barn. Here, life moved at a different pace than at the hospital. There would be no phone calls to answer until I returned home. I could watch the animals, the people, and I could listen.

Before I arrived, various household and farm items had been sold. Hand tools, horse tack, rolls of fencing, fence posts, used chain saws, and broken log chains -"Look! there's a hook at this end and a place for one at the other end." Spiles and buckets for collecting maple sap, kerosene cans, and all manner of other farmhouse staples had been exchanged and now rested in pickups parked around the yard, across the street, and along the edge of the railroad tracks.

When I arrived, several tractors were up for sale. The gray squat Ford 9N from the 1940's with one headlight pointed at the ground and a crease across its hood from something heavy and sharp went for $1,200, while the tall red International H with a modern, more powerful engine and narrow front end went for $600. "Too small for real farma and too tippy for a yuppie," someone said.

The crowd outside was given time to pick up burgers, fries, ketchup, and coffee and then moved into the barn at the urging of Herb and Larry. I stopped to look in each of the small pens near the entrance. There were pigs of various ages and types, each with a new ear tag for identification. Other pens contained geese, long necks poking from the mouth of the grain bag that was tied snugly around their necks. Newly dropped bull calves, still wet from birth, and selected for veal burger because they would never produce milk wobbled about. Sheep of various types, open faces, faces covered with wool, short legs, and long black legs crowded a couple of pens. Piles of crates used originally for shipping vegetables held rabbits, ducks, and hens. From the rear of the barn came the hysterical mooing of cows separated from their herds, the whinny of a horse, and the bray of a donkey. From the bin that held the sawdust for animal bedding rose a pine-scented aroma that blunted the odor of silage, manure, nicotine, and alcohol, brought by those who crowded the barn.

I was swept along with the crowd across the floor of the auction ring and into the bleachers. Several buyers for the packing plants and the dog food companies already occupied their usual seats and had their pads and pencils ready for recording purchases. Farmers interested to see what the market was doing, or interested in buying a heifer who would soon calve for their milking string ("She's springing, boys. Look at that bag. She's going to make milk for you!") were here and there. Thin young people in ponytails and eager expressions also came to bid on what would be stock for their back-to-nature enterprises. They were after the goats, or a pig, and occasionally, misguided, after a calf. "Gawd, that calf won't live till they get it home!" a man in overalls groaned when a smiling pair purchased a calf that had early signs of pneumonia.

Under the propane heater sat a man with a "Captain Kangaroo" mustached face and a huge belly. He hunched forward, leaning on his cane, the forward leverage of his torso partly restrained by sweat-stained suspenders connected, in turn, to a formidable seat of the pants that was anchored, in turn, to the bleacher. "LaHaye," my neighbor whispered. "A great cattle buyer, but he never buys anymore. Comes every week, though." Although the younger buyers for the packing plants and dog food companies were in modern non-descript business dress, the old-time buyers, such as LaHaye were in uniform – bib overalls, a leather wallet as large as a bible that was connected via substantial chain to a button hole, and a long animal-driving cane made from ash, the fire scorching of the crook still evident. The cane served as a prop for the chin while these men sat casting their expert eyes over feet, bags, and teats.

"Good straight back," said one of the old timers lifting his cane to point at a cow. His chin hardly moved from cane-crook height and he brought the cane back quickly to support it again. From time to time, one or two of the men arose with apparent effort, giving deep sighs, and shuffled into the sawdust-covered ring to feel a bag for firmness in one quarter - old mastitis - or thrust with the closed fist into the rear quarter of the cow to see if a calf could be "bumped."

"Now look at this!" Herb, who was in the ring awaiting some heifers, exclaimed in a voice callused by hundreds of auctions, "From up to Corinth. Man says he's short of feed for the winter and sent down these three springers!" Three large undisciplined Holsteins entered the ring. Tripping over themselves in the way that large high-school students do on their first day of work in a fast-food kitchen, the animals bungled around the ring, finally rotating in unison as the moving tips of massed canes of those at the ringside swept them clockwise. Indeed, the bags on these bred heifers were tense and bounced as the animals walked - a sure sign of imminent calving, the first for these animals, of course, followed by milk. "Boys, if I were that guy, I'd find some more hay and keep them, but this is your chance to make some milk!"

At other auctions, pedigrees, production records of dams, and a rolling herd average would be provided to guide the buyers as to the worth of the animals. At Gray's Commission Sales however, all there was to go on was the way the animals looked, the obscure reputation of a farmer 30 miles away, and the statement that the farmer had used Eastern Sires semen in his artificial insemination program. Each potential purchaser knew that "every penny used to pay for the cows would have to come out of the little hole in the end of the tit." The selling price was determined by a couple of hundred man-years of experience in estimating worth on the basis of a heifer's appearance.

As I watched the eyes of those around the ring, I thought of other situations in which this kind of expertise was required. Could bank loan officers be trained to look at potential borrowers with the same intensity that these men with overalls and thick wallets looked at the potential pay-back of these heifers? Or does a paperwork formula and numbers make expertise like this unnecessary?

Carleton began his chant, a drone of the last price and the next asking price that was punctuated by shouts from Herb who rotated in the ring with the cows, his hand pushing against the sleek side of a heifer and his eyes on the crowd for bidders. Becky, Herb's wife, freckled and red-haired, carried the clipboard with sheets on which she recorded sales, and kept an eye out for bidders also. From time to time one or the other of them would shout, "Yes!" as a bidder was identified and Carleton shifted gears to the next level of bidding.

"Are you all done at $545?" Carleton asked. "We are going to sell these cows," his voice rising in an implied threat.

"Just a minute!" Herb protested. Carleton stopped the drone dead and there was momentary silence. Herb looked sober and made eye contact with the last several bidders. "$545 and take your pick of the best of these is just not right. I said they would make milk and look at these bags, boys!" He slid his hand down the flank and under a bag to bounce it for all to see.

Carleton smiled, "Now $575," he stated confidently.

"$560."

"$560 I've got. Now $575."

"Yes! $565," Herb bellowed looking over his shoulder at no one in particular.

"$570, I've got from two," said Carleton. "Will you make it $575 sir?" For the first time during the sale he looked at a bidder.

"$575," said Carleton as the person he was looking at made an almost imperceptible sign.

"Now $580!" Herb demanded as he stood three feet from the other remaining bidder and peered directly into his face.

The person returned Herb's look with fixed expression like those at Mt. Rushmore.

"$580," repeated Herb, a bit of doubt entering his voice as he repeated the number.

"$580? $580? $580, anywhere?" Herb shouted looking over the rest of the crowd.

"Going down at $575?" there was question in Carleton's voice. "$575. Going once, Going...." thundered Carleton.

"$580, Yes!" Herb bounced on his feet, head nodding toward Mount Rushmore.

The bidding eventually finished at $590 with the buyer taking all three animals. Herb put a heavy wax crayon mark on the hip of each to document their sale. This time when the door was opened the heifers went out in single file instead of jamming through three at once - the stumble bums of fast food had completed their first dance lesson.

I walked back across the bridge after these occasional Monday evenings and along the way traded the smell of cow for the crisp freshness of the autumn wind whistling down the river valley from way up north. Overhead a million stars pierced the black crystal of night. Turning the corner, I found Polaris poised high over the smoking chimney of our house. At times I carried a pound of freshly made butter or some other small item purchased at the auction. On each trip I also carried a little bit better knowledge of cow economics.

Soon the family joined me at some of these auctions. There was a lot to learn, however, and some subjects require much patience. One night, we purchased a crate of bantam chicks, intending to let them loose in the barn where they could fly up to the rafters for a secure roosting place. They instantly contributed color and authenticity to the place, scratching in the barnyard and picking up worms and bugs. They all were given names, of course, and although I did not know one from another, the children had them all figured out. Weeks later, we began to hear a cock's crow early in the morning. At first we thought this might be coming from another farm, however, the culprit turned out to be a bantam they had named Alice. We could say in retrospect that we thought her feathers were getting very bright for a hen, but we didn't expect the early morning crow. In fact, making the call hen vs. rooster in outbred young chicks is one of the hardest calls in agriculture....until later when the rooster grows a rooster style comb and crows. Then it is easy. To survive in the business - to get to where it is easy - you must also be able to laugh at your mistakes. No event on the farm produced more smiles over the years than the naming of that rooster.

So I continued to watch my neighboring farmers and to talk with them about farming.

*****

Walter Record popped the top of the milking machine leaving the four rubber tubes and inflations behind and lugged the "kettle" to the milk dumping station. As the frothy milk was sucked off through the plastic hose toward the bulk tank in the milk room, he reattached the top with an efficient motion and slipped the inflations over the teats of the next cow in line. I found the barn a perfect temperature in my flannel shirt and light jacket. Walt was in his T-shirt and sweating with the work that flowed easily from his roly-poly bulk. Much later in his life, Walt bought some reindeer and hired out as Santa Claus during the Christmas season. There was no better belly to shake like jelly. Tragically, his career as Santa was short - his large bulk certainly a factor in his fatal heart attack.

"If you're looking for Herefords, I'd go see Shine King," said Walt in a high pitched monotone that a couple of other farmers in the town used. "He has a bull and some heifers he is looking to sell."

"I don't know Shine."

"Don't know Shine? One of the Selectmen?"

I shook my head.

"His place is north of yours on River Road. Go past the North Thetford Bridge to where the blacktop stops. Shine only ran for Selectman so he could get the road to his place paved."

I entered Shine's barn during the evening milking. A vacuum pulsated through the lines to the teats of a couple of cows hooked to the milking machine. No one was in sight. Then a door to another part of the barn opened and Shine emerged. "Checking on a cow that's calving," he said. "You here about the white face?" Shine's voice was low and harsh enough to make up for the high-pitched voices I had heard earlier - muscular, like his body. He looked at me through thick glasses.

"Yes, I'm Ross McIntyre."

He took me to the three yearling heifers and the yearling Hereford bull. They were of good size, had healthy coats of red hair, and clean white faces. There were no horns.

53

"Polled?" I asked.

"Yes, sired by a polled bull."

I was relieved. As a novice, I had no interest in being tossed into the air from the horns of an angry animal.

"$600 for the lot."

From my auction experience I could tell that this was a fair price.

"I'll take them, but I have to get the fences up first."

"O.K. but move fast. I have to get them out of here."

We rushed to get the fences up. As I put in posts, Jean followed me pounding in staples and nails holding electric fence insulators. Assisting us was a Czech economist who volunteered his labor while his wife painted scenes of our farm. He whopped away with a 24-ounce hammer, stretched barbed wire, and described his adventurous escape from behind the Iron Curtain.

By now, fall was full upon us. Late one day, I was making a gate to separate the Red Cow from the Hemlock pasture. This was the last task that was required before the pasture would be secure for cows. Looking down from the hill, I observed a truck arriving in the barnyard. Far too late to make a dash to the bottom of the hill and stop the proceedings, I realized that the truck must belong to Shine King. "Stop!" I yelled as hard as I could, but my voice was lost to the autumn breeze. Three young Herefords leapt down the ramp from the truck, cavorted briefly in the barnyard and galloped through the open gate into the alleyway leading to the pasture. They came charging up the hill while I tried to untangle the wire of the incomplete gate. I was still struggling with the wire, when they streamed through the opening into the unfenced Hemlock and disappeared into the White Gate pasture beyond. Behaving as if drawn by a magnet in Portland, Maine, they gave no indication of stopping before they encountered salt water.

I followed the cows up the hill to where their path entered the darkening woods. The sounds of breaking branches receded from me. There was no hope.

On the way down the hill, I should have been thinking of the South Dakota sheepherder who returned from the hills to the ranch 24 hours after being given his allotted thousand sheep. "I need some more sheep," he said.

"What happened to the thousand we gave you yesterday?"

"I lost them."

Instead, I was thinking black thoughts. "Damn, Shine! He could have let me know he was coming. Damn, Jean! Should have closed the gate to the alleyway! Damn New Hampshire hunting season! Begins tomorrow. I'll be up at first light to chase cows through woods where hunters will be taking 'hear shots' whenever they sense something moving. If I'm not shot, I'll then have a full day in the lab and clinic before I offer myself as a target in the twilight. $450 bucks we will never see again! Damn! Damn!" I never blamed myself for not shutting the barnyard gate. I couldn't take it. Nor did I give thanks that Shine's truck had been able to hold only three of the animals that he sold me. One remained safe at his farm.

Late that night I went to bed and tossed. Jean's breathing beside me informed me that she was fast asleep. "Such equanimity! Damn. How could she sleep in a situation like this! I guess that is why I married her. Oh, grow up! You are too mature to let a little thing like escaped cows bother you. But they will be all over town, and everyone will KNOW! How can I face the real farmers after this?"

The phone rang. I reached for it in the darkness and switched on the light, expecting that it would be a problem at the hospital. Instead it was Jean Smith. "Mike heard your cows near our house and found them. He is leading them back with a bucket of grain."

I ran outside in my pajamas to open the barnyard gate. Mike arrived just before the grain in the bucket ran out. A minute later the gate was shut and three well-behaved animals were where they should have been since late afternoon.

*****

"Time to name cows," I said at the dinner table the next night. "Our friends in Nebraska named their cows by year according to letters in the alphabet. For this year we start with names beginning with an "A". (Actually, this may not be the best way to proceed as we later learned, and some simply stick to numbers.) Putting names on them right away, however, was not only fun, but like naming parts of the farm, linked us together as a family. When we later referred to them by name, it conjured up a vision of a particular cow that all of us could share. So we began with the bull, Aloysius and two heifers, Annabel and Abigail. Six weeks later Shine delivered the third heifer, the rotund Agnes.

Eleven years later, just before we left the cow business for good, our last calf, Marimba, was born. In between, there was a kaleidoscope of names beginning with "B" to "L" as our herd expanded. As I read the pages of our farm journal, written first in my or Jean's hand and on subsequent pages with notes from Jean and the three children, I find many cows that are only dimly remembered. Some, are burned into my memory for one reason or another. None, however, are remembered as well as those first four.

Annabel was the natural leader of our small herd - the "boss" cow from the very beginning. She took the lead going through a newly opened gate, got the best place at the feeder, and, when challenged, viciously charged any bovine questioning her authority. If, when pushing the challenger backwards at full speed, the retreating cow fell, she gave the downed cow a final butt, before stalking away to eat the fallen cow's share of the feed. The other two heifers, having been vanquished by Annabel, slugged it out among themselves with Abigail the winner of that contest and Agnes drawing up the rear.

The young bull, Aloysius, was fully conditioned by his male genetics to acknowledge her authority. Hormones transform young bulls into muscular beasts that could offer a formidable contest to a boss cow. By the time the muscles of a bull take on weightlifter

56

proportions, however, they have made what psychological adjustments are required to remain aloof from cow power politics.

As Annabel grew, it was clear that mixed in with the Hereford there was a dose of "dairy". Her legs were a bit long, there was a fleck of white showing in the red coat along her back, and perhaps something in the length of her nose, revealed a bit of dairy cow. When she was bred to a Hereford bull, the calves were typical Hereford conformation, and her udder most productive. Her calves suckled fast and long, and they grew to be the strongest and largest in the herd

Annabel with her calf on her right.

The window over the kitchen sink looked out into the barnyard 100 feet away and from this vantage we became cow psychologists. When a new cow was introduced to the herd, the lowest cow in the pecking order would challenge the newcomer. This fight was not of a schoolyard type, and the rest of the cows neither took sides nor stopped their daily routine to watch. If the new cow vanquished the lowly first defender, something that seldom happened, then the next cow in the pecking order would take over and the pushing and shoving match continued, the challenger now facing a fresh defender. This process continued until, at last, the exhausted, sweat-covered, winded challenger was polished off by one of the cows in the pecking order. If, by luck the newcomer survived to challenge the boss cow, Annabel laid the exhausted and helpless beast on her backside in short order. She remained the boss. On subsequent days there were often some additional minor skirmishes in which the newcomer was finally positioned within the herd and during these days Annabel would sometimes put in her two cents worth.

It is not really a bad system for the either the cows or the farmer. When it was time to move the herd, I could pick up a pail of grain and walk toward Annabel who seemed to grasp the essence of what I was up to. We would amble toward the gate, or the door, her privilege and my security at the grain bucket uncontested by the following herd. At a gate I put the bail of the bucket over my crooked arm while I undid the chain. As the gate swung open, there was a moment when Annabel stood still and carefully scanned the green grass ahead, while the long line of cows behind her waited like cars at a red light. Then she trotted swiftly ahead, taking the very best of the tall alfalfa plants or red clover blossoms while on her way. This was followed by an uncoordinated rush of the waiting herd, heels and poop flying in all directions as the animals swept around and past their boss.

There were times when I, as the head of cancer programs in our developing medical center, sensed a similarity between my real job and what happened at the cow gate. Some of the gates I traversed at the medical center were hazardous and the territory on the far side required a careful lookover. If I had the right information and had done some groundwork the choice of the right gate could make a real difference.

*****

Boulders and ledge underlay the higher pastures on the farm. While the thin soil covering them produced a heavy growth of hay during May and June, the hot sun and dry weather of July bleached the grass and sent it into a dormant state. For this reason, we put the cows onto the hill pastures early in the year and left them there until after we had cut the first crop of hay on the better meadows. In order to accomplish this, it was best to establish a route to the hill pastures that did not traverse the lower meadows and trample the hay crop.

We chose a route that left the barnyard, crossed the end of the North Ring and then went down the alleyway leading to the Red Cow Pasture. Part way along the fence on the north side of the alleyway, I put a gate in the electric fence and cleared the brush and tree limbs so that there was a path down the bank to the brook. From there the cows crossed the brook and into the Territory a wooded area separated from the Red Cow by Jean's Stream. There was enough feed here so that the cows could forage for a week or so, and then from the Territory they could be let directly into the Hemlock pasture and areas beyond.

When this route was chosen, however, I failed to recognize the effect that the arrival of spring has on cattle that have spent all winter and then mud season in a barnyard. As the green began to show in the fields around the house, the cattle grew increasingly restless, patrolling the fence lines, and pushing their noses through the woven stock fence to gather the grass that is near the fence. It takes a strong fence to hold them in. We kept them in the barnyard allowing the grass to get a start before it was assaulted and to let the mud dry. By this time they were frantic - drawn by spring's new grass. When released into the alleyway, they surged down it, galloping toward the Red Cow pasture, until at the last second they spotted the electric fence at the end of the alley. Now there was confusion - green grass just beyond the fence, but a "knock-down" from the fence if they pushed against it. I had put the gate to the Territory, in the middle of the fence instead of near a corner and the view through it featured a steep path down across the brook, over the ledge, and into the woods - not an attractive view of succulent green.

The first year we tried this route the cows ran back and forth in the alley never spotting the exit to the Territory. I grabbed a bucket of grain and led Annabel to the gate, down the stream bank and into the woods. The other cows followed. I had failed to recognize until this moment, however, that the hierarchical system governing cow behavior does not extend to calves. They follow their mothers but do not do so blindly. They remained behind, terrified of the narrow gate in the electric fence, and as yet, uneducated to the advantages of following the older animals through it. The cows, having had an appetizer of the grass in the Territory began to respond to the cries for help from the calves, and some of them came charging back through the gate. In the confusion that followed, a calf bumped the electric fence, snorted, bucked and butted another calf and a melee followed. I refilled the grain bucket and tried again. After a lot of effort, eventually cows and calves arrived in the Territory. Each year we went through this, and each year we were always too busy to change the geometry of our fences and gates, although, in the middle of the melee I always swore that I would take care of it the next day.

One year, no matter how much grain I used, I was unable to get all the herd beyond the gate to the Territory. One rambunctious calf remained in the alley, and I gave up, figuring that once it got hungry enough for milk it would go through the gate. Anyway, it was getting late, and at 6 AM the next morning I would have to be at the airport for a flight to New York. I had a late dinner, checked and found the calf still in the alley, and then Jean and I headed for bed. My suitcase was packed, the shower was comforting, and I lay down to the bawling of the calf in the distance. It continued, now with a cow in answer, as I drifted in and out of sleep. I sensed that the sound from the calf was becoming more hysterical, and noted that the sound was now coming from several different directions. I could imagine the calf, having left the area where the electric fence separated it from its mother, was running back and forth along the stock fence. The mother, too, as I awoke again, sounded more concerned. Then there was the sound of wire squeaking over staples. I tried to ignore what they were doing to the fence and closed my eyes again. A mistake! This was followed by the sound of cedar fence posts snapping and wire being dragged.

I struggled awake and into my farm clothes. In the mudroom I grabbed the big flashlight, and I picked up the lariat from the barn as I headed out to the alley. Fifty feet of fence was on the ground. The calf and cow stood on either side of the mostly downed fence. In the dark and with the calf distracted, I threw the lariat. Amazingly, it looped her head and I pulled it tight. By the time of this episode, I had used the lariat enough to know that I could get the calf to the gate. Bit by bit, snubbing the lariat around trees and posts whenever the calf tried to run away from the gate, we inched toward my goal. Finally, having got myself through the gate and using the help furnished by gravity as I stood on the steep stream bank, I yanked that calf down the hill and into, if you will, its mothers arms. We were not done however. I had to get the lariat off the calf, which by now was emitting choking noises. This lariat was a modern version in which the burner, the leather-bound eye that holds the noose, is replaced by a steel burner that undoes with a catch. This greatly simplifies the release of a struggling animal. I snubbed my end of the rope, went hand over hand to the calf, as do the ropers in the rodeo ring, and released the burner.

The calf took one look at me as her brain received the first real dose of oxygen in some time. She had not spotted her mother, and bounded up the hill toward the gate. My steps matched hers and we met at the gate. It was almost a fair fight, but I had decided that this was one I couldn't lose. I got an arm around the neck, grabbed a leg and together we rolled over and over down the bank and into the stream. The comings and goings of the day had left the track well greased and it was a fast trip for us both through mud and cow manure. We lay together in the water for a bit and then she got up and walked over to her mother. I picked myself up, coiled the lariat, climbed the bank, closed the gate and went in search of my flashlight. It was then I realized that it was pouring rain. I remember thinking, "I wonder how long this has been going on?" and also, "You and this farm are absolutely crazy!"

Later that same day in New York, I sat at the magnificent table in the boardroom of the Mount Sinai Hospital and wondered if the people in adjacent seats picked up any of the aroma that the second shower of the night had not removed. I also noted that the captains of industry and finance, various former trustees of this great institution whose portraits graced the walls, appeared to be looking at me critically. Perhaps even they had some inkling of what I had been up to. It didn't bother me much. I was trying too hard to stay awake.

*****

We were finally at a point where we could start to sell some of the herd each year. The steers and the cows we culled were taken to Sharon Beef in Vermont. A week later Jean would return to pick up a truckload of old liquor boxes containing the nicely wrapped and frozen beef and set out on a delivery route to those who had purchased a quarter or half when it was "on the hoof."

Other animals we sold as yearlings, and sometimes as a cow and calf pair. Of course, in order to sell the good animals, you have to keep and breed at least some of the best. The market for frozen and wrapped beef takes care of those that don't answer either need.

One year we sold a fine cow and a rapidly growing calf to a dairy farmer in Grafton who wanted to start a beef herd. He and his son arrived in their truck on a Saturday afternoon. With little difficulty we loaded the cow and calf and they were off. By late in the evening, however, it was clear that a mistake had been made. We had a bawling calf and a mooing cow that were not a pair, and we assumed that the dairy farmer had noted the same problem. We had loaded the right cow but the wrong calf. I had been especially busy that spring and had not come to know my calves as well as I should have.

Since it was late and dairy farmers go to bed early, I decided to wait until morning to call.

At 5 AM I called, "This is Ross McIntyre and I'm sorry to say I helped you load the wrong calf."

A woman's voice responded. It was edgy, "You are sorry, and so are we. That calf jumped out of the barn through a window and is running through the corn leaving a trail of blood. All the men are out there looking for it."

"I'll load the right calf and be down as soon as I can."

Young Ross was up early. He and I put the sideboards on the truck, got the cattle into the barnyard, caught the calf (by this time in the story we had built a chute into which the herd could be run and easily separate the animal of interest) and loaded it. It took about an hour to get to the farm and by that time, the bloodied and frantic calf had been caught. The corn looked nearly as bad as the calf.

It took only a few minutes to make the switch, and Ross and I departed leaving a contented calf/cow pair. The situation in the back of the truck was quite different, however. The calf deprived of its mother, hungry, and bleeding from cuts inflicted by the broken window glass, was by no means content. It strained at its halter, butting and kicking the sideboards while offering hysterical cries - now faced with another uncertain destination. We pulled out onto the highway.

"Dad, what is Eastman?" We were passing a large sign directing people down an attractive road.

"That is a four-season second home community that is getting underway."

"Is that the one with a ski hill?"

"Yes, they have a golf course, beach, and a ski development."

"Could we get a look at the ski lift?"

"Sure."

We turned down the road past the immaculately tended golf course bordered by well-maintained yards and gardens of proper vacation houses as we cruised to the bottom of the ski slope. By now it was 7:30AM on Sunday morning. Suburban Eastman, just awakening, was getting a strong dose of lost calf as it kicked at the slats and bawled. From under the tailgate oozed a liquid that left an unmistakable trail on the new blacktop. To those who live there I now offer my apologies. I really felt that I owed this to Ross who had arisen early and worked hard to rectify my mistake of the day before.

*****

To detect when a cow is in heat, those who tend cows are forced to observe cow behavior as a clue to this event. Some people are good at this, and some are not. A bull has absolutely no problem in making this call. A person may note that every 19-20 days a cow gets twitchy and that other members of the herd may ride it, but the suspect cow also rides others. Just which animal in this riding event is the one in heat is baffling to an amateur.

The bull uses another method. From time to time he sidles up to the rear of a cow that has just unleashed a yellow waterfall of urine, catches a last drop from the vulva and snorts it into his nostril. With his nose extended into the air and with exquisite concentration on the deep breath that he is taking, his nervous system, acting as a biological gas chromatograph, makes the determination. As simple as that, he either walks away, or jostles the cow's flank, the first step in foreplay. At times the cow holds back a good dollop of urine to the very end of voiding - a situation that leaves the bull with urine running down his face and a surprised, sometimes incredulous, look in his eyes. Fence-sitting children love this kind of accident.

The heat often lasts less than 24 hours so you had better be there to observe it if you are going to use artificial breeding. We decided that with our schedules, we would not know our animals well enough to make artificial breeding successful. That was the reason for Aloysius, our first bull. With her eye on the barnyard from the kitchen window Jean could see him sniff, stretch, and make his determination. We wanted to be certain that he had bred the heifers, however. There is certainly no profit in feeding heifers all winter if they are sterile. So a couple of months later, I had Ed Blaisdel, the vet. come down to Lyme from North Haverhill to check things out.

It was a good visit. I got the cows locked in the stanchions before he pulled into the driveway. As he drew on his boots we traded on the common currency of medicine - animal and human. We understood each other immediately. He wanted to know about whether we had found the old dump for the house, and I knew that he was interested in antique bottles. I told him about the bottle of Warner's Safe Liver and Kidney Cure, a brown cast bottle bearing an impressive relief of a safe that we had already found.

He pulled a long plastic sleeve and glove over his right hand and arm, soaped it, and inserted it far into the rectum of the first heifer. "Yep, she's pregnant!" He was feeling not for the embryo in the uterus but for the presence of the corpus luteum on the ovary - the tumor-like mass of cells responsible for the secretion of hormones required to sustain the pregnancy - and he had found it. Annabel and Agnes were pregnant. Abigail was not. We got the bill later. For his long trip to Lyme, for standing on slippery boards next to rambunctious cows, and inserting his arm into them up to his biceps, he charged us a pittance. Such is the world of large animal vets. Wonderful people!

We gave Abigail a bit longer to be bred, but it didn't happen. She had to go. But now it was mud season and the barnyard was too soft to get the truck in. The children had each adopted a cow by then, and when the weather was sunny, they petted and played with the cows as they lay on the bedding and chewed their cud. Elizabeth, who was now six years old, had adopted Abigail and spent hours sitting on her, from time to time riding on her back as she arose and walked off. It was going to be a tough separation.

When the barnyard was finally dry enough, I caught Abigail in one of the end stanchions and erected a fence of 2x4s next to her. When I had finished this, I revved up the Squire's truck and with spinning tires churned backwards up to the platform where she stood imprisoned. We were set for the morning.

I arose early, ate breakfast, carried my office clothes out to the truck, put some grain in the truck bed, attached a halter to Abigail and looped it over the irons that held the front of the truck body together. Keeping tension on the halter, I released the stanchions. This was the first time that an 800-pound animal and I had been on the same rope together. With a magnificent jerk of her head, the rope burned through my hands. This was followed by splintering of the 2x4s as she leapt sideways and cavorted away. From along the fence came a child's voice, "Run Abigail!"

A quick glance showed tears mixed with delight on Elizabeth's face. I was now off chasing the loose halter as it snapped over the ground. Soon the barnyard was chaos – with lots of hooves and flying manure. I got the end of the rope around a post and bit by bit regained the stanchions. I rebuilt my fence now with stronger planks, and this time she peacefully entered the truck.

I set off for the Sharon Beef Company dripping oil from the leaking main rear bearing as I drove, checking my watch frequently. Cow day at Sharon and Hematology Clinic were always on the same day. I had a large list of patients beginning at 8 AM. It was going to be very close. The unloading was quick and in the tiny rest room at the slaughterhouse, no larger than Clark Kent's phone booth, I unrolled my office clothes exchanging them for my filthy farm clothes without misfortune. A few minutes later I was in my white coat, tie, and greeting the first patients. I felt like Superman, but it had been a close call. I also had to face Elizabeth at the end of the day and I didn't feel like Superman then. We had a solemn discussion about the economics of farming and sterile animals. She was calm, gracious, and understanding, but terribly hurt. The resolution of our conflicting goals that day took place via a conduit of good will, forgiveness, and love. You have to load a child's cow to fully understand it.

*****

One evening in the fall of our second year on the farm, Jean and I climbed up through the Potato meadow to check the water tank. It was dusk and the light was fading fast as we climbed. Sal, our Labrador, slunk off to one side of us. He was avoiding the herd that munched peacefully. The water level was fine and we returned down the hill. As we approached the fence, Sal, for some reason, made a playful run at the bull, pulling aside at the last minute as she approached him. Aloysius did not take this as a friendly gesture and exploded after her. Seeing that she was already well on her way to the fence, and frustrated by his inability to send her flying, he fixed on us.

I had thought about the possibility of being chased by a bull from time to time, as all people who own bulls should, and had developed various contingency plans. Included among these, as a last resort, was the bullfighter approach - some sort of cape and fancy stepping coupled with an "Ole!" When it happens however, such fantasy doesn't enter one's mind. The bull comes on like a truck, and not one that is out of control either. It has precision guidance, adjusting perfectly to your attempted evasion. We sprinted toward the fence, separating a bit so that he couldn't kill us both at the same time. (A good marriage is when such actions occur without discussion.) As we dove under the fence to the sound of ripping shirts we considered the possibility that the wire would not stop him. But it did. Only then did the rush produced by our squeezed adrenal glands finally arrive to leave us excited, trembling, and with our hair standing on end. Plans to dispose of a bull are easily made when hormone levels are high.

Having taken care of Aloysius, the next January I followed up on a notice in the Market Bulletin for a bull that was for sale in Newport, NH. I turned off the highway toward a house that had a striking barn beside it. It looked a bit like winter quarters for a circus - perhaps for the elephants - Victorian decoration, huge doors, fine hardware, but now worn and weary. Who could have afforded this fancy barn on their farm? A herd of Herefords lounged under the barn and in the adjacent barnyard. Among them was a fine young bull. I appreciated the honest assessment of the bull that the owner, Mr. Sichol had given me over the phone. Indeed, this bull was as advertised. We looked over the herd and I asked him about the barn.

"Oh, the barn was built when they brought the buffalo to Corbin Park."

Corbin Park was a huge fenced game preserve on nearby Croydon Mountain - formerly a playground for the ultra-wealthy, and now for the merely wealthy. For a time it had been used as a site for restoration of the buffalo herd after its near extinction in the 1800s. The barn would have held enough hay for a lot of buffalo.

At the end of the month I returned and picked up the young bull, Norman Domino. That load, probably the heaviest the Squire's truck had ever carried, put us on our way in the cattle business. Not only did he have a good pedigree, he never chased us. In March of 1973 Annabel had a calf by Norman Domino. It looked good and grew fast. Early on we decided he had the makings of a wonderful sire. Since we were in the "D" year for calf names and because we saw the promise of this animal, he deserved an appropriate name. He was christened "Dingdong" one night as we sat at the dinner table under the hanging Sears Roebuck lamp. The name was perfect for reasons that we all knew and understood. It required no discussion. There were no secret snickers, because there was no secret. We seized on the name.

We sold Dingdong from our herd before he reached breeding age, because we wished to keep Annabel and didn't want a mother-son breeding problem. After the sale we heard tales of his accomplishments from up and down the valley, often from people who had no idea where the animal, (or his name) had come from.

Norman grew into a formidable animal but by 1973 his offspring had matured to the point where he could breed them. He had to go. So there was a sad day when we collected $575 from the buyer ($350 more than he had cost) handed over his papers and loaded him up. His genes remained behind in his offspring and we felt good about that.

In order to locate another herd sire, I went to some other Hereford operations but didn't find what I was looking for. Then I followed up on an ad in the Market Bulletin from a place on the other side of the state. The farm had an impressive sounding name and I imagined as I drove over there that I might encounter serpentine walls leading up to a mansion overlooking Lake Winnipesaukee. Instead a dusty drive led up to buildings that were clearly the home of a shoestring operation. An old-timer named Fred emerged and showed me the herd. It was in good shape. He mentioned that his partner, Tom, was a policeman from a town in Massachusetts and that the two of them shared the management of the herd. They had an artificially bred registered bull that looked good. As I left, Fred told me that Tom would call me to discuss terms.

When the call came, I was surprised to learn that they might be interested in loaning me the use of a bull in return for his keep over the winter. First they wanted to look over our place and our cattle. They arrived a few days later and we took them out to see the herd. We had just let our cows into the heavy new growth of the replanted Potato Pasture. As we stood waist deep in a tangle of sweet smelling red clover and grass, the heads of timothy just emerging, each mouthful from that field was about 25% protein. Tom was a huge guy, with neck like a professional wrestler, and a bald head that needed frequent mopping in the June sun. He smiled and responded in a voice much too small for his frame, "This sure looks good to us!"

My experience with policemen up to that point had been largely one of being intimidated. The incidents were minor - I had crossed over a freshly painted highway center line as I dodged a pot hole and left ugly paint tracks as well as tell-tales on my tires, or I had forgotten to register my vehicle and been caught, etc. Each time, I had sat anxiously in the driver's seat watching the approach of the officer in the mirror, knowing I was guilty. Now I was standing in a meadow getting ready to do business with a cop who was as big as any two of those I had encountered previously. I wondered what would happen if I took the bull and it sickened while on my hands, or if it ran through our fence and got hit by a car, or how I would like telling this huge fellow that lightning had struck his bull. His fascination with the healthy growth in the pasture and his approval of our animals calmed my fears, and the deal went through. I hired Gerald Hughes to take his cattle truck over to pick up the bull and "Creator", too large for our pickup, rode to our farm in style. He was a good choice, and he stayed healthy. Later, after Creator was returned to them, Tom and Fred loaned us a second bull, "Collegiate" whom we nicknamed, "Joe College".

The naming fun continued. Lady, who had been bred to Norman Domino and had a calf we named Chandelle (or candle) in the "C" year. Chandelle was bred to Creator and her calf in "D" year was "DeLight" and in the "E" year, "Eveready". Her calf in the "F" year was Flame, whereas DeLight's calf that year was Flicker. So it went around the kitchen table at naming time - better than anagrams. For a while we all could keep them straight. Then, after we sold the wrong calf, we got ear tags.

*****

Before leaving the cow business, I have to relate one more incident. Agnes was rotund when we bought her and became fatter month by month. It was striking. No other cow came close. I wish from the opportunity I had to watch her grow, I could provide some information on the cause of obesity, a controversial subject, to put it mildly, but I can't. It is with brutal honesty that I say I don't know whether Agnes ate more than the others, whether she exercised less, or whether she ate the same amount as the others but used it more efficiently to make fat. But fat she was.

One winter we arrived home in the late afternoon and we all went out immediately to do chores before dinner. It was apparent from extra space at the feeder that a cow was missing and it took only a moment to deduce that it was Agnes. I found her in a corner of the cow barn, up against the old whitewashed walls. Trapped on her back! Her four legs pointed in the air, her eyeballs rolled. Her huge protuberant belly was pressing on the diaphragm. She gasped tiny breaths!

Not a moment to lose, "Help!" I tried to roll her but she was well stuck and weighed 1000 pounds. The kids arrived and we all pulled. Nothing. We put a rope around her legs, then leaned on it. Couldn't move her. We didn't have a come-a-long and the rope on the block and tackle was broken.

"Ross, the tractor!" He went running while I opened the gate for him to enter the barnyard. The tractor starter groaned a few times and died. No charge in the battery. Jean arrived. "Get the station wagon and back it into the barnyard. But don't come so far you get stuck in the mud. I grabbed a heavy manila rope and tied it to Agnes's far leg. I ran the rope across the room, out the door and through the barnyard. Jean was just backing through the gate while the rest of the cows watched - fascinated by the unusual activity. I put a loop around the trailer hitch on the station wagon.

71

I didn't want Jean to apply power slowly, bog down and stall the car in the mud. "Giv'er plenty of gas!" She did. She floored it!

The rope twanged tight with a tremendous jerk and then went slack. Jean stopped within 10 feet. I ran into the room expecting to find a 3 legged Agnes on her back and the fourth leg, pulled from its socket, still hooked to the end of the rope. Instead, Agnes was standing on four legs and belching like Vesuvius did before Pompeii. We cleaned up and went to dinner where the conversation was entirely devoted to complements on Jean's driving skill. She had known that she had one chance and she wasn't going to lose it.

The Herefords line up for a health inspection. Some hay with a drizzle of molasses brought them up to the fence. When these cows had calved we peaked at 27 head of Herefords, all that the place could carry with our six horses and 20plus sheep.

Chapter 6

Horses

I may have given the impression that cows were the most important animals on the farm. To me, they were, but if it hadn't been for Jean's interest in horses, we would not have moved to the farm. And she could not have persuaded me to start the search for a farm if she had not detected, and exploited within me, a trickle of interest in horses.

This trickle did not include the horse show business. It goes back to the 1940 riding class (military hands and seat) my brothers and I attended at one of the few livery stables at that time in Omaha. The instructor, Captain Plumb, was a no nonsense survivor of a few years in the U.S. Cavalry. He did his best to teach us how to ride a horse - tearing his hair when we didn't get it right, frequently - and becoming violent when one of the numerous plugs from the livery misbehaved. There was only one really good horse in the string, and that horse was assigned to the oldest and best customers, who of course, won ribbons in the shows. The rest of us struggled with spiritless, or recalcitrant, or absent- minded nags, and the boredom of endless drills in the ring. So much for horse shows.

We all would have dropped out of class early but for the few trail rides we took through Elmwood Park. On these, the excitement furnished by Plumb's response to the messes we got the horses into kept us interested. For instance, when a horse lay down, discharging its rider and then proceeded to roll over on its saddle while the helpless ex-rider stood in the bridle path holding the reins, it was clear we had trouble in the making. Watching this insolence continue, the ex-rider's face now puckered in pre tears, it made the whole class feel better to hear Plumb's U.S. Cavalry bellow, "Colummmmn Halt!" and to witness his horse galloping back down the column. Seeing the big man hand his reins to the sniveling youngster while taking those of the offender was a good first step, while kicking the horse upright and then springing into the saddle was inspiring. And then - he had spurs and we did not - planting those into the flank while he brought the horse into a full gallop, now, that was retribution of the first order. Just before he disappeared in the distance he did a sliding stop, putting the haunches nearly on the ground, and returned even faster than he departed. By this time, the bounce was pretty well out of the horse. He and his student would exchange reins once more, and our ride could continue.

Though there may have been some in the class who fantasized about taking over for Plumb in these situations, I don't think there were many. I know my own interest in horses was directed away from the class and the ring, toward a subject I could deal with only in the obscure corners of my mind. It centered on the Palomino that I had observed a couple of years earlier in the parade during "Golden Spike Days". This was an event engineered by Cecil B. DeMille to celebrate the premier of his movie, *Union Pacific*. Not quite 75 years after the pounding of the Golden Spike at the completion of the Union Pacific Railway, most bars in Omaha, the home of Union Pacific's home office, put up a log front and installed swinging doors. Men changed their denim overalls and their hammer loops for real "cowboy" jeans. I heard the word "Levis" for the first time.

My mother cut up an old leather jacket to make me a vest with a leather fringe. Dressed in it, I stood at the curb with my family watching the parade and awaiting the passing of the convertible carrying Barbara Stanwick, the star of the film.

The street was full of horses and wagons, floats, clowns, funny cars, and bands, but my attention was drawn to a Palomino that cruised along the street next to our curb. Its heavy saddle was bound in silver with chains of silver medallions on the rigging, and it came along, slanting into the crowd while doing a pleasing cross step, as if to say, "photograph me". In the saddle, was a trim blonde woman, perfectly dressed for the occasion. She handled the reins with a lightness that would have made the bit acceptable to any of the young men along the parade route. She was beautiful, on a beautiful horse, on a grand day, and in an era before the term "eye contact" was used, I did my best to make some. Her eyes, however, were directed out over the crowd to no one in particular or to all in general, and with a few more cross steps she was gone. The vision remained, however. I wanted one of each: the horse, the silver, but, especially, the blond woman. I was 7 years old.

She had held the reins so gently in her left hand and waved with the right to the crowd. The horse responded instantly to the light pressure of the rein across the neck. As we plodded around the ring under Bob Plumb's direction, the vision of that Palomino, seen two years before, must have been lingering in the subconscious. In our military stance, we held two reins in each hand and pulled alternately on one or the other handful of leather as if they were coupled to the steering bar on a rusty bobsled. In my fantasies as we circled, I imagined being on the Palomino, holding the reins in one hand lightly across the neck and having the horse respond to the command. I sensed the power and acceleration, the muscle underneath, that would be available as a magic carpet to unexplored places - an animal that would rejoice with me as we sped along, our heads drawing closer together as I leaned forward over the neck in the wind we were making.

Then, several years later, to my surprise, I found myself aboard a horse like that - one that would respond to a gentle touch, one that moved as if it were connected directly to my brain, one that loved to run for the fun of it. Almost 20 years later, I found another, but it was too much to hope that I would ever own one. To protect myself from disappointment, I buried that dream. In the background, though, the beautiful woman on the horse remained - just a blur in the corner of my eye, out of focus, only a distraction if I let it be one, but enough at last when the time came so that Jean got her farm and her horse.

Also, out of the corner of my youthful eye, I saw and learned things that later were to prove helpful on our farm. I watched a beautiful quarter horse unloaded from its fancy enclosed trailer at the livery stable after a trip up from Texas. Purchased for the princely sum of $1000 in 1941, it must have had the traits that I was dreaming of. The proud owner jogged it on the trail alongside the road to warm it up, and then let it out for a brief canter. The military hands and seat class watched enviously. To think that that fine horse would be boarded in our livery stable with the old plugs! Two weeks later the quarter horse was dead. "Shipping fever." The plugs remained pictures of health. The word from the hangers-on around the stable was, "If they had gone to Texas, paid $100 for a horse, and hauled it back here through rain and cold in an open trailer, it would still be alive! But pay $1000 for a fancy horse that's been washed and combed to within an inch of its life, it is going to die even if you brought it up here in a Pullman."

*****

Jean did not get her Palomino right away. Nor did she ever get a silver-studded saddle. We worked into the horse business slowly, making at least as many mistakes as we did in the cow business. As we moved along, however, we got closer and closer to making my fantasy real, and on special occasions, from high in the saddle, Jean would make eye contact with me.

Jean and Mike's wife, Jean, started wintering camp horses at the Smith farm a couple of years before we bought our farm. Many summer camps for children have a string of horses that are used for about 8 weeks in the summer. For the remaining 10 months of the year these animals must be fed and housed. In order to avoid this cost, some camps offer "free" horses to those who will care for them. For the 10 months, the person with the borrowed horse is not only responsible for feed and shelter, but also for routine health care and shoeing. The middle of winter is not an ideal time to go riding in New England, so this "free " horse is less of a bargain that it might be otherwise. With luck, the arrangement amounts to 4 to 5 months of good riding weather.

The two Jeans had approached a nearby camp about getting camp horses for the winter and the camp was glad to oblige. The woman who owned the camp had two horses delivered to the Smith's farm. It was a good introduction to horses, since these were old, stiff, and relatively bombproof. The horses decided during the first few days that they might be spared of any inconvenience if they refused to accept a halter or bridle. This was an excellent tactic, and the acting, ears back, snorting, prancing, and the like was flawless. It certainly flummoxed the two women for several days. The horses, however, had not counted on my concern that we were going to be feeding them for 10 months during which time they might not be ridden. It was this economic threat more than any bravery on my part that forced me into the low shed next to the Smith barn, where I cornered one of them and put on its halter. The other, sensing that the game was over, was easy. The two Jeans witnessing this capture were impressed but had no idea that my success was entirely cheap-skate driven.

The farrier was called in early spring to shoe these horses and the two women asked him about the other horses at the camp. "Too bad that she wouldn't let you have Ginger. Now that's a horse! If (the camp owner) was as good looking as Ginger, she would have been married long ago."

The next year two horses from a camp with an elite riding program spent the winter at Mike and Jean's place. One of them, Princekin, arrived at our farm a few weeks after we moved in.

This was at the time that the cows were eating our pastures down fast, and we were putting up new fence in every spare moment, getting a new field ready for the cows just as they ran out of forage on the one they had been in. To relieve the strain on the available pasturage, Jean bought a kit that was comprised of an auger connected to a swivel and some rope. When screwed into the ground, the auger was anchored. Attaching one end of rope to the swivel and the other to an animal's halter would, in theory, allow the animal to walk in circles while grazing in an unfenced field. The apparatus was referred to as a "calf stake" since it was frequently used to tie out calves. Some people use them for horses. I hope they have more success than we did.

As soon as she got it home, Jean screwed the auger into the ground near the grass and clover about 50 feet to the rear of the mudroom door and tied Princekin to it. For a few moments, all went well. It was evening and what had been a gentle breeze was changing to something a bit more boisterous - a weather front just arriving - the gelding's coat being furrowed a bit by the increasing wind. At that moment, Ross, eager to find his mother instantly for some reason, threw open the screen door on the mudroom, ran into the yard, and shouted, "Mom!" The spring on the screen door, having never been stretched quite this far by its previous owner, was assisted by the rising wind, and the two conspired to slam the door with a force that rattled the glass in the nearby windows. The loud "Mom" was followed instantly by a formidable "BANG!"

Princekin, not quite certain what to make of his new surroundings, put on edge by the approaching weather, and restrained by something designed to cope with a 100 pound calf, not a fully grown horse, reacted instantly. Dirt flew from the auger as it popped from the ground, and he disappeared around the corner of the house dragging the rope and the calf stake. I ran around the house following him only to see his rear end speeding south on River Road, the vicious looking auger clanking on the pavement like the chain under a gasoline truck.

I filled a grain bucket and hopped into the truck. By the time I got to the Four Corners where River Road and the East Thetford Road cross, there was no sign that Princekin had passed. I stopped and got out to look for hoof prints, signs of scraped pavement, any clue I could find. It was clear that he had made it to the crossing, but try as I might, I couldn't deduce which of the three possible choices he had made at the intersection. Crushed at my inability to track what should be something easy to follow, I returned home, and called the police to report that there was a horse loose on the road, and "by-the-way, he is tied to a calf stake that will disappoint anyone who happens to run over it!" I then went out again and searched all three roads, but saw nothing.

We had a sleepless night. Early the next morning the phone rang. It was Myrtle Bailey, "Mrs. McIntyre, are you missing a horse? We have one on a rope wrapped around our diesel tank."

Jean and the children went over to a farm about two miles away and found Princekin, rope tangled in the legs of the tank. They recovered the horse, the rope, and the calf stake. There was no fire, no one had driven over the stake, and the horse was uninjured. If we had paid a thousand dollars for him, he would have been dead.

Later that fall we learned that the secretary for the Department of Medicine was eager to sell her horse, a nondescript black mare, Sheba. Jean drove out to Lyme Center to ride her and returned with the news that she was a solid horse, gentle, but neither well trained, or well-gaited. However, the $250 price was right, and it was alleged that the horse was trained to drive. By November 23, we had two horses but no good place to put them.

The horse barn needed serious work before we could lodge the horses, and the end of November is none too soon to be finding a stall for a horse in this part of the country. A well-fed horse with a good winter coat does fine in the snow as long as the wind isn't blowing hard, but a cold rain will cause hypothermia. So the horses went into the barnyard where they could share an open bay in the cow barn with the rightful inhabitants. Better housing would have to wait. I had been named as a Markel Scholar, a nice honor for a young academic physician, and the annual meeting of the Markel Scholars was scheduled down south. Spouses were expected to participate in the meeting. Jean's parents arrived to take care of the children while we were away and off we went.

When we returned a few days later we found Jean's parents a bit worse for wear. In addition to all the usual tasks which they so gladly performed, there were some that were unexpected. I don't think that we had told them that Princekin had developed the habit of pounding the galvanized steel barnyard gate with his hoof to call attention to his desire for his morning can of grain. This resulted in a daily metallic serenade at about 5:30AM that served as our alarm clock. Her parents, being new to the situation and somewhat slower in their morning routine, kept him waiting for longer than was usual. By the time her father arrived carrying the can of grain, Princekin had moved laterally from the gate and had pounded the fence, instead. The heavy-duty woven wire fence had a bottom strand of 10-gauge wire that was now lodged between the horseshoe and the hoof, trapping his hoof in the fence. Fortunately the horse, having more intelligence than many of his kind, had stopped pawing and stood at attention as Jean's father approached.

As a young man Jean's dad had wildcatted oil wells in the back woods of Kentucky with a drilling rig pulled by teams of mules. With this experience for background, he stepped into the barnyard and addressed the problem before the horse tore down the fence and became entangled. No matter how he tried, however, the hoof remained trapped. As he worked, Princekin gave a sudden lurch, tearing the shoe from the hoof, and destroying the section of fence, but not becoming tangled in it. Now this kind, elderly man had to rebuild the fence on a cold end-of-November morning before he had his breakfast. As he described this event, I felt like a heel. I still do. We were overextended, unprepared, undercapitalized and irresponsible. The only difference between our affairs on the farm and my work at the medical center at that time was that in the medical center I was not irresponsible.

While Jean's folks were still there, I needed to build a half a dozen wooden gates, each 12 feet wide. I used some of the Squire's hemlock boards. After driving the tractor out, Jean's father and I laid them out on the floor of the hay barn where the tractor had stood. I cut, he nailed, and the sawdust flew. We were rolling creosote onto the last gate when Jean drove into the driveway in the Squire's pickup. She parked the truck and walked up the ramp to the barn slowly, observing the stack of newly minted gates. She spoke rather deliberately, "The truck rolled over at the Penfields."

"How could it? You just drove into the driveway."

"George Lawson came with his tow truck and pulled it up. It seems OK."

She had left the truck parked in the driveway while she and her mother went into the Penfield's house. The Penfield's driveway sloped steeply to the road. Across the road was a deep ditch, and a row of trees, with a field beyond. The parking brake on the truck was of doubtful ability and we always put it in gear when parking on a hill. Once before when I had left it parked this way, it gave a "chuff" and then another "chuff" and then a "chuff, chuff, chuff" as the force of gravity overcame the leaky compression in the old engine. With each chuff, the truck made a short hop in the direction that gravity was pulling it. Fortunately, I had not made my exit and could use the foot brake to stop it. Since then I had adopted an inconsistent routine of putting chocks under the wheel when parked in steep places.

Jean and her mother had been chatting with Abby Penfield when her husband, Don, entered the room, face ashen, hands trembling. "Was anyone hurt? Are you OK?"

"Why, what is the matter?"

"Your truck turned over in the ditch!"

As the three of them left the house and stared down the driveway, disbelief was replaced by a mental replay of the truck chuffing its way down across the road, into the ditch, and slowly rolling over.

George Lawson owned the local garage. It was housed in a recycled metal building that he had moved to a site on Route 10 just north of the village. His parts room consisted of a fine collection of junked vehicles that lay quietly and unobtrusively behind the garage. He claimed that his most important tools were his screwdriver (a formidable brute about 24 inches long), a hammer, and a crowbar of the sort that railroad tracklayers used. He did have a good set of wrenches, but he did everything he could to avoid using them. His old red Ford wrecker had a stout derrick on it. When he was called for a tow, he sped through the town with his unmuffled exhaust barking through a straight pipe that served to advertise that he had been called and would be out of the garage for a bit. As he flashed by, one could catch a glimpse of him hunched over the steering wheel, his thin face blending like that of a raptor into the renegade image he deliberately cultivated.

George arrived at the Penfield's within a few minutes after he was called, positioned his truck, ran the cable to a solid structure on the Chevy and pulled it right side up. The gas filler pipe, which was located on the passenger side to the rear of the door, had been pushed into the cab during the rollover. George lay down on the seat, braced his shoulders against the driver's door, and pushed the

82

filler pipe back into position with his foot. I still have his bill for this complete service: $10.00. Jean was back on the road.

Taking note of the deliberate manner in which Jean had delivered the news, I calmly finished the creosoting before I went to look over our little truck. By this tactic, I attempted to hide my concern about the spilled battery acid that I imagined was right now eating its way through important structures, the loss of coolant from the radiator, and other sorry outcomes of this event. Her dad and I calmly surveyed the sheet metal. As in the case of horses, if I had paid big money for the truck, the outcome would have been different, but with the Squire's truck there was virtually no evidence of its recent encounter with gravity. The sheet metal, radiator and battery were fine.

The horse barn with the cow barn and corn crib behind it. The hooves drummed on the new floor. The end of the cow barn pokes out behind it and the corn crib is further on the right.

I hired a Dartmouth student who had worked in construction during his summer vacations to help me get the horse barn ready for occupancy. Together, we muscled the remnants of the old floor out of the stalls. The nails were rusted in place by their years of exposure to damp manure and most had to be broken in order to lift the old planks. We jammed a large crowbar between two layers of floorboards, and put our weight on the other end of the lever. The nails broke with a pop, sending a cloud of dried manure and chaff into the air, our clothes, our hair, and our noses. The student was a good guy and must have needed the money because he agreed to come back on future weekends.

Eventually, we had a huge pile of debris that we hauled to the slope behind the corncrib and dumped down the bank. The remaining solid skeleton on the west side of the horse barn was ready for reconstruction. We patched the sub floor where necessary and put down a layer of tarred-felt on top of it to keep it dry. The new hemlock planks were so full of sap that when they were nailed the hammer splat droplets as the nail drew tight.

Nailed tight, the hemlock dried in place, straight, level and much tougher and more resistant to abrasion than the pine we had replaced. We fixed the broken boards in the mangers, put some iron over the surfaces susceptible to destruction by horses teeth (known as "cribbing"), laid in a supply of sawdust, and hauled in a couple of pickup loads of hay from Shine King's supply. Finally, we were ready to lead the horses in. The new floor, tight like a drumhead, reverberated with a solid beat to the horses shoes, a sound I remember with pleasure and sense of accomplishment.

*****

With few exceptions, one should never go to an auction with the idea of purchasing a specific item. An open mind and a broad knowledge of what things are worth are helpful. I've said that the livestock auction represented the convening of a willing seller and a willing buyer. A really good buy at an auction, however, occurs when the seller is not entirely willing. It may be like real estate, where a friend has told me that, "You never sell land you always lose it. You never buy land, you always take it off someone's hands." Our first pony was a good buy.

One night all five of us sat in the auction bleachers being entertained while learning what cattle sold for. At one point the ring was cleared of a group of rambunctious cows and a solitary black Shetland pony was led into the auction ring. Before she was auctioned, however, Carleton and Herb sold some lead ropes with bright snap fasteners on the end. A price was established and multiple buyers held up their hands when the auctioneer offered additional ropes at the same price. As these were being tossed from the ring to people in the back of the bleachers the pony stood amidst the flying ropes – for all the pony knew these could have been flying snakes. The contrast between the cavorting cows, flying lead ropes and the pony was remarkable. Released now from the grip of the handler, the pony stood calmly inspecting all portions of the room and assessing the place. Her demeanor clearly indicated that she was not intimidated by the situation she found herself in, nor was she because of her small size, inferior to anything she caught within her gaze. I almost heard her say, "I know who I am, and who you are. I am independent, free, and equal to the best of anything that has ever come through this ring." She wasn't bragging or arrogant, she was just telling the truth.

The pony's calm intelligence was offset by her appearance. She had a good coat of winter hair, but it was marred where large patches that had fallen off - tufts of loose and falling hair surrounded these patches. Something was wrong - lice or worse. A human with such wretched looks would have been on the defensive, embarrassed to be there, seeking to hide. Instead the little pony wore her impairment almost defiantly, perhaps akin to an abused woman showing her bruises as she seeks a restraining order in court. Jean and I exchanged glances; I felt the eyes of three children upon me.

Carleton began the chant, "Nice little pony. Who'll gimmee a hundred, hundred, hund..."

No need to be coy here. I called out, "Twenty-five!"

"I've got twenty-five, now make it fifty, fifty...... Who'll make it thirty five, thirty five, thirty five?"

There were no other bids.

A cheap little saddle, probably consigned with the pony, came up soon after. We bought it too.

It was a euphoric group that looped a piece of hay rope through the halter and led the pony out across Route 5. The tiny hooves clipped along on the blacktop, the stars danced, and we were headed for home. There was no jerking on the rope, the pony knew she was going - as one of the gang and she scarcely had to be led. As we approached the house I remembered the can of flea powder on the shelf near the basement stairs. We had used it months ago to treat a dog. We stopped outside the horse barn and doused her with it before she could infect Sheba or Princekin. Then she was taken in to one of the tie stalls, a little wisp in a stall intended for a plow horse.

The children named her Cinderella (Cindy), and the effects of the flea powder on her coat were like a touch from the wand of a fairy godmother. Within weeks her coat was beautiful, but the arrival of beauty did not change her personality - she remained simply one of the gang. The saddle went nearly unused as the children hopped aboard bareback. She was completely devoid of the stubborn streak found in so many Shetland ponies. In fact, she was like a Labrador Retriever that could be ridden. The children trotted along the woods roads, cantered in the fields, lay over her back like a sack of grain, chased loose cows on her, fed her garden produce (including, I think, zucchini), and lived with her.

Later, when large horses occupied the four tie stalls, we moved her to a small stall I built at the north end of the barn. I found a fancy headboard from a bed and incorporated it into this accom- modation. This unusual feature reflected the capable femininity of the occupant, and Jeanie hung over it a smooth piece of pine bear- ing the name, Cinderella, in Magic Marker calligraphy.

This little pony provided the best evidence that Jean and I had that our children were growing. Month by month their feet came closer to the ground when they were on the pony's back. Elizabeth who was only five years old when we got Cindy, had several years of good riding, but eventually, even her feet got too close to the ground. By then there were other ponies and horses to offset the loss of this dear friend and we sold her to a family in Norwich. It has been more than thirty-five years since she went to join those new children. As you can tell, however, I have not forgotten that she knew who she was, and how to have fun.

Elizabeth and I staged this photo for a Christmas Card. Her pony, Thistle is happy for a jaunt in the snow.

*****

In contrast to the self-confident and optimistic outlook of Cindy, I felt that Sheba displayed the limited outlook that I associate with downtrodden employees. I once watched the facial expression on crowds of women leaving a large shoe factory where they had spent all day stitching and gluing countless pairs of boots. Their future held only the promise of more of the same tomorrow. Whether that future is what dulls the mind or whether it is dullness that puts one in the position of having to do more of the same tomorrow is a long debate. Certainly, as events proved, I was not one who would change Sheba's view of the world.

As I overlooked the many deficiencies of the Squire's pickup truck and used it for all that it was worth, so Jean overlooked Sheba's deficiencies. She was delighted to have a young horse with some bounce in its legs. Princekin was getting stiffer each month - less and less able to carry a rider. Sheba was learning a bit each month and growing more enjoyable as she learned. I suspect that Jean disagreed with my assessment of Sheba's personality. Jean was a very forgiving person.

Shortly after Sheba arrived, Jean had gone down to a house on Shoestrap Road and returned with a sulky she had bought, a light rig with two bicycle type wheels. One of the two bright red shafts had cracked part way through, but the price was right: $25. I had some fiberglass and resin for repairing rusted fenders and I used it for a crude but effective repair. When finished, the split shaft was bumpy but the red painted wood peering up through the lumps of fiberglass and resin was stronger than when it was new.

By then, Dick Jenks had smoothed up the North Ring with his bulldozer and we had put up a rail fence to encourage the horses to move in a circle. Jean walked Sheba around in this confined area with the horse hitched to the sulky until she felt that it was safe for the next step. Now it was time for her to sit on the tiny seat with her legs spread-eagled around the rear of the horse and each of her feet cupped in a footrest alongside a rear leg of the horse.

All went well, and within a few days Jean made her first tentative trips out onto River Road. She moved off smoothly with a gentle "clip clop", well-oiled wheels turning without apparent effort, and disappeared down the road. I felt anxious about this driving venture, however, and I believe that the children shared my worry. A short time earlier, Dudley Johnson, a neighbor, had purchased a fancy driving pony and pretty four-wheel cart. His pony moved out smartly and he cut a fine figure trotting over the town roads. One day, however, his pony had bolted, taking Dudley on a careening ride until the pony and cart left the road, bounced across the road-side ditch and into the Connecticut River. Dudley bailed out just as the cart sank. Despite his efforts to save it, the pony, tangled in the harness, drowned.

If anything, the sulky would be harder to bail out of than a four-wheel cart. So I worried. My ability to worry is something that Jean and I joked about for all the years of our marriage. The success of our relationship was in large measure our ability to find humor in what should be serious situations and to her ability to relieve my anxieties - to take the unexpected in stride. I directed. She floated. She knew a lot more about horses than I did, as will become even clearer as this tale progresses. This was her horse and her sulky. It was hard for me to deal with it, but she was going to trot. I couldn't stop her and I had just enough sense not to try.

Jean and her horse Sheba arrive back from a trip on River Road during the peak of fall colors

Also about this time, a childhood friend of Jean's delivered a very nice stock saddle as a gift to us. It had previously been used on a horse on a New Jersey farm and that horse, now ancient, was no longer ridden. Now that we had a saddle that would accommodate a rope, I had a vision of working with Sheba to see if she might be useful when it came time to move cows. I knew enough to rule out any possibility of turning her into a cutting horse, but I thought that perhaps it might be possible to rope a calf from that saddle. First of all, I had to get aboard.

Jean and I put the new saddle on her. I think Sheba was surprised by the unusual feel of this western saddle. It would have been best to walk the horse for a bit while carrying the new saddle as an introduction. Instead, I put my left foot in the stirrup and swung up. At least I had the sense to pull a bit on the right rein while I was on the way up as she wheeled to the left carrying the saddle away from me. That was, however, the only thing I did right that day.

Within seconds she reared, then again, and again, each time higher. I thought of one of my mentors in hematology who was recovering from a crushed pelvis after a horse in Texas had reared and gone over on him. Higher and higher we went until I decided it was a good time to slip off. As soon as I stood on the ground beside her, we were friends again.

It was time for a consultation with my friend, Will, a technician in the Pathology Department. He maintained a smoothly functioning autopsy room and in his spare time told stories about his horses. I found him sitting on a tall stainless steel stool, his muscular arms and hands engaged in cleaning a sink full of instruments.

"Will, what do you do with a horse that rears?"

"Tell me what happened."

I told him.

"Well, one thing you can do, is when the horse starts up, reach forward and give the horse a rap between the ears with your fist." He demonstrated, putting aside his cleaning, and leaning forward off the stool, gave the table in front of him a good thump with his clenched fist.

"I never thought of that one!"

"Well, try it."

The next weekend, we put the stock saddle on again. This time I made it out onto River Road before Sheba gave a tentative little rear. Following Will's advice I gave her a solid rap between the ears. She went down and then reared again. I gave her a bigger rap. She went down and then started up again. I gave her all I had. She went down. As she started up again, I had the feeling that this time it was all the way over. I bailed out, much less successfully than the week before, landing on my ass on the blacktop. A car that had slowed so that those within could watch the proceedings inched past me as I sat on the road, holding the reins of the now calm horse. Inside were two of my medical residents and their wives. Humiliation! I caught the look of surprise on their faces as they recognized the blacktop-sitter and wished that I had possessed the presence of mind to cover the lower half of my face with a bandanna, bank robber style.

Spring came and Jean called the director of the summer camp to remind him of the arrangements made the previous fall concerning the return transportation of Princekin. The director decided that it was not worth sending a trailer all the way across Vermont to pick up his aging horse and told Jean that she now owned it. This was the first time we realized that anyone with a farm is at risk of becoming an assisted living facility for aging horses. From that time on our growing stock of horses included two types: young relatively untrained animals, and senior citizens that we had a hard time refusing as gifts. There were few in the middle.

*****

During our second summer on the farm, Jeanie used her baby-sitting money to purchase a Welsh pony from Charlie Hewes' farm on the Goose Pond Road. The Hewes had a large string of ponies that they took to the various country fairs where they set up a riding ring for youngsters. Jeanie had worked with them in the riding ring: selling tickets, adjusting stirrups, and sleeping out with the rest of the pony ride crew in the back of their horse truck when she had to stay over the night. Curtsy was a friendly black pony with a back broad enough for the kids to stand on while riding.

A short time later, Ross went to Gray's auction and bought a large pinto pony. She got the name Omaha, to reflect her western heritage and now he was suitably mounted. His friend, Steve Wurster, also obtained a pony of similar capability and the two of them embarked on what became nearly daily expeditions to the forest and the hills. If Ross was not helping me make hay, he could be found on his pony. Jean took a picture of Ross standing next to Omaha, one of the few that she had time to process in our bathroom darkroom, during those hectic years.

The horse barn still had room, and although the children were delighted with their mounts, we could see the time when each child would be moving up to a larger horse. Furthermore it was exciting to contemplate additions to our stable. Jean found that Wayne and Vi Wilmot, who lived a couple of miles south of us on River Road, had a grade Morgan stallion and that the stud fee was fifty dollars. Even better, we didn't have to get a trailer to move Sheba to him. Jean simply got on, rode her down the road, and brought her back, bred - like in the old days.

It went so well, that Jeanie took Curtsy down the road also. We were now set for two foals in June of 1972.

June arrived along with tropical storm Agnes. Each year I took two weeks of precious vacation during the middle of June in order to get the first crop of hay in. With Agnes' arrival and her aftermath, I watched each day pass while sodden hay rotted on the ground. Halfway through my two weeks we had just 450 heavy wet bales in the barn. To prevent spontaneous combustion, I cut the strings on the bales and mixed a hundred pounds of salt with the hay. Then the storm took an unexpected gyration and dumped rain on new swaths I had just cut. I stood in the field and shook my fist at the sky. It made me feel better.

Meantime, Jean and the children were watching the expectant mares. When the farrier put shoes on Omaha that spring, he called attention to a fetus kicking within her, too. We had not two, but three foals coming! We concluded that a stud at the auction had bred her - the dates were right - and the package she carried was going to be a surprise since the stud could have been anything from a racehorse to a donkey.

Our mares were put into the Red Cow meadow, where there was plenty of room and good forage. Jean and the children made frequent visits during the day and went out through the darkness in the middle of the night with flashlights to check on things. Soon it became clear that something was dreadfully wrong with Omaha.

Her belly began to swell in directions unrelated to her pregnancy and her hindquarters became swollen. We removed her from the pasture and put her into the box stall in the open bay of the cow barn. Ed Blaisdell, the vet, came down and checked her over, but could not make a diagnosis. We kept her confined and watched her grow slowly weaker.

Jean looked out the window one morning and saw Curtsy trotting back and forth along the fence line. Unusual behavior requires a check. She and the children found all the horses in the northwest corner of the field. Here we had put the electric fence several feet back from the steep bank of the pasture that dropped down to the brook. Just beyond it were the remnants of two previous fences, both entwined in the trunks of large trees. One of these was rusted barbed wire, some of which we had been able to clean up. The other was a much more formidable stretch of woven wire fence, now collapsed upon itself, embedded in tree trunks and slumping down the bank, some of it stretched like a hammock over the brook itself.

They searched for the cause of the mare's discomfiture and found it in the shallows of the brook. Curtsy's new foal had staggered, rolled, fallen not only through the electric fence but also through the life-catching potential of the other fences and landed in the small stream. She was alive, and it was with triumph that Jean and her small helpers retrieved her and delivered her to Curtsy.

The next night, Jean and Jeanie went out at 2AM with their flashlights to check on Sheba. The tumultuous weather of tropical storm Agnes continued. In the distance there were flashes of lightning and the rumble of thunder. The horses were not found immediately, and while searching along the stone wall near the Hemlock Pasture the storm was suddenly upon them. The lightning flashes and the thunder came nearly simultaneously, and a burst of cold wind drove a deluge whose streaks bent back the beam from their flashlight. They were up near the cow-scratching tree, the huge bull pine - so-called because it had never been cut - with the horizontal limb polished by innumerable cow backs stretching out toward the southwest. We are all told, of course, never to seek shelter under a tree when caught in a thunderstorm. We are also told that you shouldn't be in the open, especially if you are the highest thing around. So what do you do when you are under the trees and the only shelter is 400 yards away across an open field? Do you stay or do you run?

They ran. As I think back on Jean's reentry to our bed that night, her toweled hair and head regaining the pillow that she had abandoned just before the storm arrived, I realize that the ambition that drove our farming operation achieved its momentum by spreading us too thin, by putting us alone when we should have been together. There was no comfort, no sharing, on that stormy hillside.

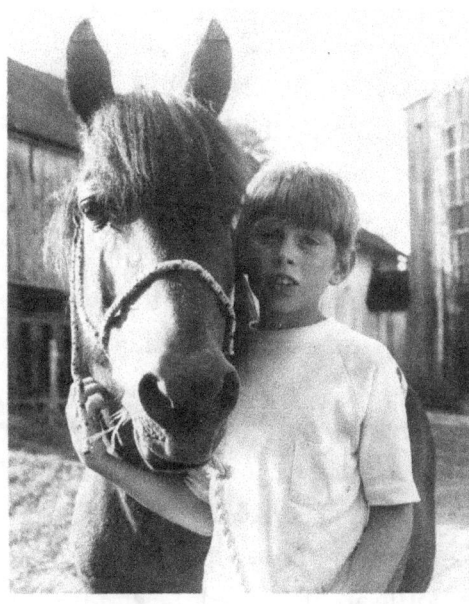

Ross and his pony Omaha, best friends

Meanwhile, Omaha continued her downhill course. I found the whites of her eyes gone yellow with jaundice and wondered whether the abdominal swelling and edema represented a liver problem - as an oncologist I had to consider cancer. She became weaker by the hour. As evening came I took Jeanie to a concert at Dartmouth for which we had purchased tickets several weeks earlier. Jean, Ross, and Elizabeth would stay with Omaha. A short time after we left, the pony went into labor and delivered her foal. Then she began to stagger. Jean called Ed Blaisdell who could advise only that no one should enter her stall since she could fall on them. In a short time Omaha's

95

head dropped to the bedding and she died. She had never been ridden with her new shoes.

We arrived home to find Jean, Ross and Elizabeth with the aftermath. The foal had been dried and had been given a bottle according to Ed's suggestions. They had also been told that the chances of raising a healthy orphan were not at all good. Omaha's lovely pinto body lay out in the bay of the cow barn. The next day was a big day for me at the hospital. Now it was time to assign turns for those checking on the foal and Sheba, and to go to bed. I walked upstairs and entered the bedroom. There on the mantle piece was Jean's photo of Ross standing next to Omaha, his small hand curled under her chin and around the halter. I saw the expression of love on Ross' face and for the first time that night fully recognized the extent of his loss. I broke down - still weeping as I turned in. A short time later, Jean climbed in beside me. She was crying too.

In the midst of the continuing storms, Sheba now delivered her colt - lively in good health, no fence rolling, no problems. Ed's assistant came down the next day and we did an autopsy in Omaha's stall. She had died of a ruptured uterine artery. It had been leaking slowly for several days, producing a huge hematoma that was responsible for her abdominal swelling, the edema and the jaundice. When the foal was delivered, the pressure holding back the leak was gone and she had died of a rapid hemorrhage.

Ross wrote the entry for her birth in the farm journal: "Nebraska's Stormy was born. The mother was Omaha." Reading that note today makes my eyes water still. The other foals romped in the Red Cow meadow, while Stormy took her bottle and followed the children as surrogates for her mother. There was another matter to attend to.

Buried under other more recent notices on our bulletin board in the pantry was a sticky label. On it was a drawing of a Holstein cow with all four feet in air and flies buzzing around. "Wanted: Down and Dead Horses and Cows Maxham Fur Farms, Inc. Worcester, Vermont 223-6335 - 7 Day pick-up service. *Cash for prompt calls.*"

The knacker is an indispensable member of rural society. The ridicule of the people who do it by those who don't have to do it is a well-known phenomenon in farming cultures. The jokes told about knackers are as old as they are tasteless. The knacker's best recourse is to approach the job with humor as our long-time knacker did. The "Fur Farm" was a figment going back many years to the time when there actually was a fur farm.

So it was that the knacker came to our farm. He pulled into the driveway with his truck, took one look at the sea of mud between the gate to the barnyard and the stall where Omaha's body lay, and pulled out his long steel cable. He slogged through the mud, made the cable fast to Omaha, and returned to his truck. Slowly the winch took the pony across the barnyard. Its' passing left a muddy ditch that filled with water. When at last the pinto hide and contents had slipped up the ramp and the tailgate was closed, we were left only with the ditch pointing north as evidence of Omaha's passing.

The birth of Stormy made the loss of Omaha a bit easier for us all to deal with. This foal required the full attention of the children. They used a large whisk to stir the milk replacer powder in a bowl of warm water, poured the liquid into a robust gallon plastic bottle with a red rubber nipple, to be greeted by the mouse colored foal at the barnyard gate. She sucked up gallons of milk replacer and grew stronger day by day despite Ed Blaisdell's dire prognosis. Between bottles was a time for hands-on love, grooming, and other motherly tasks. She played with the other two foals, kicking up her heels and bucking across the pasture. When the other foals went off to be with their mothers, Stormy, at a loss, found a child for food and comfort. I believe that she survived entirely because of the interest and attention she received from the children.

About this time Jean's mother arrived from New Jersey in her tan Dodge sedan and parked it next to the vegetable garden. Stormy, being led by a child past this parking area, suddenly lay down next to the car; its still warm engine making gentle clicking noises as it cooled. Finally, she had found a large warm object as a mother - life was good! While Jean and her mother weeded the asparagus, Stormy lay for hours next to the Dodge. I thought for a time we would have to switch cars with Jean's mom when the time came for her to leave, but soon Stormy would accept any parked vehicle as a surrogate for her missing mother.

The other two foals of that June should not be forgotten. Thistle, Curtsy's filly, never looked back after being rescued from the brook as a newborn. She soon gave evidence that she was to grow larger than her mother and that the choice for her name was right on target. She was not ornery, just "prickly", when from time to time she was asked to do something new or different. In my preoccupation with our orphan and many other matters, I paid little attention to her. Jeanie, Curtsy's owner, took full responsibility for training her foal, until one day three years later I realized that she and Jeanie were nearly fully grown.

I hoped that Sheba's colt would turn out to be the kind of horse that I had always dreamed of riding. Jean named him "Sheik". I liked his looks, and he grew into a handsome bay with white stockings.

If any of the suicidal depression that came to a peak in the Steele family of 1931 lingered in the fabric of our farm home, it was exorcised during 1972. Our foals kicked up their heels, playing "chicken" in near misses, farting, and spreading joy all around. In character, I tended to spread worry about injuries and threats of grief, but my effort was unsustainable. The foals encountered fences and other hazards, tearing patches or chunks of skin from vital areas. I gave awful warnings which were rendered ridiculous by treatment outcomes. Love, soap and water, a few bandages, and these wounds healed as if they had been skin grafted by the world's best surgeon.

A neighbor, Gordon Heard, a breeder of Welsh Cobs used for pulling, had an attractive mare that had accidentally been bred by an Arabian stallion. He gave the resulting cross breed to Jeanie. Jeanie named her for one of the constellations, going down the list of those mythic heavenly constructs until she came to the word, Carina, a constellation in the southern sky. The word derives from the Latin word for keel, a part of the boat that confers strength and stability.

It was a good name. She turned into a beautiful large pony - almost a small horse. Well-behaved, intelligent, and friendly she was a perfect companion for our eldest child. She also provided us with an introduction to Gordon Heard and his pony farm.

Heard, an older man with a twinkle in his eyes, presided serenely over the chaos resulting from the amalgamation of pony farming and care for several men who in today's politically correct syntax would be referred to as "educationally challenged". The children may have occasionally used an earlier term, "feeble minded", but most of the time simply referred to them as "Gordon's Men".

Heard had recognized that a small pony farm was a perfect place for such men – a place where they could be assigned simple tasks in a country environment supplying good food and plenty of exercise. Although not bright, most of the men had the good sense to refuse to stack hay on hot days, leaving the task to others. The exercise Gordon prescribed for his men daily came from walks along the highways in Lyme. Depending upon how rambunctious the men were on a particular day, Gordon would drive a shorter or longer distance down the highway before letting them out of the car to walk back to the farm. By the time they returned to the farm the steam was pretty well out of them, and they were easy for Gordon to manage. Gordon's men could be recognized easily because they were the only fellows to be found walking along the road in sports jackets and ties.

This civilized road-walking dress led some children in the Lyme School to spread a rumor about them that the teachers were unable to suppress despite their best efforts to do so. The theory held that the well-dressed men had, at one time, been brilliant students in

Jeanie levels her eyes with those of her pony Carina. My fantasy of a walk to the Pacific Ocean with the pony was kindled by a book I read as a boy, Stephen Meader's *Boy with a Pack.*

school but that they had studied too hard and had "burned their brains out."

Their dress also sometimes caused problems for others. One day Ross and his friends were fishing in Post Pond near the bushes alongside Route 10 when they heard a car brake to a stop. A voice from within the car called to several of Gordon's men who happened to be walking by at that moment. As Ross and his friends hunkered in the bushes and struggled to contain their laughter, they heard the person ask for directions in a down-country accent. When the car drove off, its occupants even more confused than before they had stopped, the boys imagined what the people were thinking. "If the well-dressed people in Lyme are as dumb as that, I wonder what the ordinary citizens are like."

So it was wonderful to have this careful small Carina who looked after herself, to remind us of Gordon's pony and man farm. She became my favorite. She led so easily and did so much to please that I had a recurrent fantasy of taking a long pack trip with her. Across the country? Why not? I studied the simple weathered wood of a crossbuck pack-saddle in a museum of Native American crafts and imagined sculpting some wood to make my own saddle, cutting the rawhide lacing from an uncured hide bought at Gray's auction. Escapist? Yes. Just what I needed when times got tough at my real job in the Medical Center.

During the years at our first house in Lyme we had watched Glenn Buzzell training horses. He lived on Market Street in a house that lacked an adjoining pasture. However, like many houses constructed in the days before autos arrived in the town, there were horse stalls in a barn next to the house. Although part of the barn had been converted to a garage, Glenn had enough space to stable the horse that he was training for someone else. Glenn was a retired sawyer, the person who looks at the log on the carriage of a sawmill and decides where to make the saw cuts in order to maximize the fraction of it that can be converted to merchantable lumber. A good sawyer may get one more plank out of a log than a poor sawyer, and this plank, multiplied by all the logs going to the mill, may spell the difference between economic success and a failure for the mill owner.

Glenn selected his words as carefully as he selected his saw cuts. The result was an economy of words used and a maximum of impact, so appreciated that each time he rose to speak at a Town Meeting, we all recognized that we were in for a treat, no matter what position he took on the item under debate. His approach to horse training was similar. We saw him walking past our house with a horse on lead while using his analytical mind to anticipate and react to each move of the horse, but with such restraint that he appeared on the surface to be simply walking beside the horse.

We tried to emulate him, but it was not easy. I suspect that a really good horse trainer is a person who can exclude all extraneous thought and sensory inputs while with the horse. Instead, as we walked beside our horses we were thinking of the things we had to do when we and the horse got back to the barn, planning the next day's work, contemplating schoolwork, the next budget committee meeting, or what to serve in the next meals on wheels. We spent a fraction of our day with the horses and the horses learned from us. But we never had lesson plans, or a time set in the calendar for teaching, and we lacked the rigor of formal testing let alone certification. If, as is said, one must build three houses before getting one that is right, then why should the first horse that one trains be perfect?

Such was the case with Sheik. Why we never asked Glenn to do the training, I don't know. This term introduces us to the topic of education, which I don't intend to dwell on, except to say that there are those who believe that one can define an educated person as one who has attended school, and others who believe that you define an educated person as one who has learned something. If a horse has been sent to a trainer, you may put an advertisement in the newspaper describing the horse as trained. Whether that horse will stop running after it throws you may or may not have been in the curriculum. Sheik was trained, but the curriculum was limited. Though many of Jean's training objectives were never achieved, one memorable success stands out. She trained Sheik to stand quietly beside the mailbox while she reached down, opened the door, and removed the mail – the mailbox was not available from the backs of our other horses.

Horses bring different innate skills to the school of life-long learning. Some never learn to pick up their feet as well as others, some find fences to cut themselves on while others never tangle, some run harder and buck more after stepping in a yellow-jacket nest, some are dull and some are high strung. There is a vast difference in the behavior of "cold blooded" work horses and "hot blooded" race horses, the genetics of breeding for these purposes influencing what these breeds can be taught and how they will behave in a given situation. Whether one is killed by a kick in the face by a Percheron, or tossed headfirst onto a rock by a runaway Thoroughbred, the outcome of bad training or bad genetics is the same. The selection and breeding of horses by humans that has given us our present diversity of breeds, attributes, and skills has been, in large measure, driven by the need to select desirable traits while at the same time minimizing the possibility that we will be killed while taking advantage of these traits.

Although I have emphasized genetics, I don't deny the importance of non-genetically determined events. When Sheik was a yearling, Ed Blaisdell paid a visit during a Saturday morning when I could get free. We led Sheik out onto the grass in the North Ring, gave him a shot of tranquilizer and stood by as he weakened and eventually collapsed on the turf. Ed had me sit on Sheik's head while he maneuvered Sheik's scrotum to a position where he could apply the clamp to the horse's spermatic cords. Neither Sheik nor I were unscathed. I don't know whether the trauma of that event got through the tranquilizer effect to permanently imprint on Sheik's psyche, but I sensed that our relationship was never quite the same after that Saturday. Perhaps it was just my belief that the relationship had changed, and Sheik thereafter sensed my belief instead of my weight on his head.

Educational philosophy is not something that occupies the front burner in one's day-to-day life with horses, however. One has to set aside a special time or be long finished with horses to reflect on and ponder the question as to why breeding is so important to what a horse can be taught. Is it intelligence that is inherited, or is it temperament? Is the ill-tempered horse dumb because with an ill temper it cannot be taught? Or is it ill tempered because it is dumb? How much of temperament is heritable? How much is learned? Why is it that eye contact is so easy with some, and so hard with others? It is during the time of this reflection that questions flow as to why so little attention is paid to the impact of genetics on the education of humans.

Instead, during one's day-to-day life with horses we react emotionally. While walking from the gate to the barn with a pony ambling beside me, the loose lead rope there for symbolic purposes only, I gaze deeply into the blackness of its eye wondering if I am indeed, connecting through neural pathways that eventually arrive in the pony's cortex. The pony, appreciative of the handful of grain that awaits it in the barn, in turn looks into the smaller blackness of my pupil. I know it is making a connection with my frontal lobes. I feel comfortable. I have respect. I love.

The horses found the spring-fresh sap trickling from a wound in the large beautifully shaped maple that lay on the fence line separating the Red Cow from the Hemlock Pasture. In a few days the bark was completely stripped and this fine tree, the bearer of autumn's best color, is consigned to firewood. I'm angry. I could kill. But the fault was mine. I should have fenced off the tree.

We sometimes cover up our emotional response to the horse's misdeeds, since we recognize that we have to take responsibility for many of them. Jean and the children are better at covering up than I am. One day Jean walked from River Road into the driveway leading Sheik who was pulling the sulky. From the look on her face and the perspiration it had been a long walk. Something was wrong, but Jean didn't say much. A week or two later, she sold the sulky. I never got the details of what had actually transpired during her outing, but it was significant.

One evening, I was up in the north corner of the Red Cow fixing the electric fence that had shorted out. Looking down across the Home Meadow I could see the children, the ponies, and Sheik gathered around the hitching rail near the horse barn. Suddenly, something was wrong, Sheik was bucking and Jean was on the ground. I ran the 400 yards and arrived in the midst of confusion. Jean lay in the driveway, seriously hurt, Sheik was still bouncing about so I shouted and waved at him to get him out of the picture, throwing my hat at him and suggesting that I would kill him as well. Elizabeth didn't calm me a bit, when she ordered, "Forget the horse. Take care of your wife!"

Jean had landed on her knee. She was in great pain and was now growing a bit cold and clammy - the initial stage of shock. In a few minutes we had blankets, a pillow splint and she was stable enough to be lifted into the car for the trip to the emergency room. The bones were all right, but when the swelling had subsided the orthopedist removed the damaged pieces of cartilage. "Looked like crabmeat", he said - cartilage that was certainly missed later.

A few days later I noted that the wobbly chair with the fiber seat bottom which we sat on while changing boots had been moved over against the wall cabinet in the mud room. "Why was the chair there, instead of its usual place?" I thought. "Someone must have moved it there to get something off the top of the cabinet. And what is up there? Ah Ha! The ammunition for the rifle!" While I was at the emergency room with Jean, the children had removed the ammunition to a safe place in case I intended to make good on my threat against Sheik.

How angry does one have to be in order to resort to mayhem? I never intended to kill Sheik; it was an idle threat made in the heat of the moment. But, if Jean had been killed, what would I have done? The reaction of the infuriated male of the human species to a horse in a situation like this may be an ingredient in the genetic selection of horses. It may be responsible for the proposition that we are less likely to be killed riding horses now than when man first climbed aboard a prehistory horse. Killing a gelding because it has injured someone does not, of course, result in improving the genetics of the breed. It may only prevent further injuries. From my childhood I remember the horse, Spider, who went on injuring people time and time again, until, finally, someone put a stop to it. It should be clear by now that Sheik and I spent no time gazing into each other's eyes. My fantasy of riding with the wind on a beautiful horse that we had trained never materialized.

The Palomino? Well this gift arrived at our assisted living facility for used-up horses while I was at work. Jean had me go out to the barn and there he was, a real surprise. A nice looking horse, too. Perhaps this was the end of my quest. I took him out into the North Ring and we warmed up a bit before going to an easy canter. In the turn he stumbled a bit. Just enough, so that when the stumble was repeated a few more times I knew that we would never be looking for a silver studded saddle. He did its best and we had fun with him for a while, but it wasn't right and we knew it. The horse is gone, and so is the name. It didn't last.

I don't want anyone reading this to think that the horses of their fantasies are really beyond reach - they are out there - just hard to find. While we were in the late stages of the horse business, I went out to Dallas for the annual meeting of the American Society of Hematology. One evening of this meeting is set aside for a reception and relaxation. The Program Committee of the Society had taken a gamble and arranged a party and rodeo at a dude ranch instead of the usual hotel ballroom affair. I admired the spirit of the Committee, although what they arranged turned out so badly that the Society will be holding their reception in ballrooms for the next century.

The "dude ranch" was not a dude ranch at all. It was a place to which Gray Line tour buses brought tourists, mostly from overseas, to see a completely staged representation of Texas ranch life. It was set up to accommodate several hundred guests for a couple of hours while they watched a truncated rodeo and ate some barbecued beef. The Society had just begun a spectacular growth in membership and instead of the fifteen hundred expected attendees about three thousand showed up for the meeting. City buses, packed solid, were used to shuttle the members to the "ranch" where they were disgorged onto the hopelessly inadequate infrastructure of the place. The "rodeo" commenced with the small bleachers at the edge of the ring bending under the strain while hundreds of hematologists pushed their way to the ringside, only to find that the drinks were elsewhere. They gave no attention to the weak public address system on which the poor announcer was reciting his usual Gray Line text concerning the fine points of roping a calf, while a couple of forlorn cowboys on fine horses tried to key their moves to the announcer's text. It was like trying to give a lecture on gravity to students on the beach at spring break.

I looked around and observed two fire pits. Over each a very small steer was roasting and I decided that the chances of getting a square meal were poor. Buses continued to disgorge their loads until there was scarcely room to move. The management, realizing that disaster was near, pulled those putting on the rodeo from the ring and assigned them tasks in crowd control. The dinner gong rang loudly followed by a rush that swamped the staff and the food. Bill Grace, who was a trainee in our program at the time, and I watched the sharks depart for the food. I knew he could ride and so I motioned at the two roping horses that stood calmly, their reins on the ground, or "ground tied" awaiting the return of their riders. Given the mob, their riders might never return.

Bill and I climbed the fence, approached the horses, which accepted us into their saddles as if we belonged there. The ring had been freshly dragged to give a soft but secure footing, and the horses behaved like the ones in my fantasy. Behind the backs of three thousand people headed to dinner we turned, we wove, we galloped, we changed leads with the smoothness of a fine automatic transmission - it was perfect. The effect of the reins on the neck was like having my own legs extended to where the horses touched the ground. I thought of where the horse should go and it was there, like magic. We missed dinner, but it was better than any dinner, and certainly better than the one the rest of the Society had. It proved that the right horses are there. You just have to find them.

Of all the photos taken on the farm this is the one that carries the most meaning for me. The heavy new snow had silenced the world about us, and although I wished that the horses would gallop through it as I held the camera, they had no intention of doing so. They were right. That would have broken the silence and it was better to have that done softly by the cascade of snow from a pine tree. I stood with this thought: What I saw was attributable to the thousand and one things we had all done in order to make this photo possible, and how unlikely it was that this moment would ever repeat again.

Chapter 7

Dogs and Chickens

When our poor Labrador, Sal, had to be put down because of cancer, we took a pup from a bitch owned by the Lyme minister's family. The bitch was a good dog, but had been bred by a dog of uncertain origin that had paused as he walked by. Don't ever take a pup from a litter where the sire is "just on his way by".

Vincent was a bad dog. Named after Van Gogh, his personality, when matured, enabled a life-style with an uncanny resemblance to that of his namesake. If the dog, Vincent, had possessed thumbs he would have cut off his ear. Unable to paint, he expressed his creativity by digging, effectively and with bold strokes. Into these holes he buried treasure that some day will astound a searcher with a metal detector.

No fun to walk or play with, Vincent chewed at his doghouse until any engineer would have declared it hazardous for occupancy. Straining against the chain he dragged his disintegrating abode in and out of the several open pit mines he had dug. To diminish the damage we let him run free, and he then wandered over the hillsides inflicting mayhem without guilt or remorse. The Smiths started to miss their bantam chickens. One by one they disappeared. When Vincent was caught, I explained to the Smiths that Vincent took only those chickens that were slow or roosted too close to the ground. I was stalling for time. I didn't want to do what was obviously necessary.

Vincent finished the Smiths' bantams and then started on ours. Alice vanished. "O.K. that's it!" I think Jean and I uttered the words simultaneously. Vincent, unwisely, had rejected all traits that would have produced allies amongst the children. The decision was easy: Herb Perkins.

Herb was Lyme's Dog Officer. He was also Lyme's Constable, weather prophet, and cider maker. He raised hogs for us and others in the town, and recounted the history of Lyme to those who assisted as he cut pork chops for our freezers. He performed all these jobs with a grace that flowed easily from his bulky figure. With so many friends in town he almost never switched on the blue gumball light that perched on the roof of his aging Plymouth Valiant. He left traffic enforcement to those with more capable vehicles and spent his hours as an officer directing traffic for weddings and funerals.

Jean put Vincent in our truck and drove him up Baker Hill Road and on to Herb's house, on Isaac Perkins Road. His house, a lovely unpainted Georgian, was crowded by trees growing close to the foundation. Herb's house had been there a very long time.

She pulled in past the cider mill and met Herb out back. Vincent hopped out eagerly - places where pigs are butchered and dogs are kept are of particular interest to other dogs, especially ones like Vincent. While there, Jean looked at two dogs that Herb offered her as replacements, a perky beagle and a larger quieter white mongrel.

When Jean returned she described the two replacement possibilities while we ate dinner. I could sense that the children were imagining the dogs in various locations around the farm and trying to decide which would make a better companion on the basis of the limited information Jean could offer. It was a toss up. Either would be O.K.

The next day Jean called Herb. "Do you still have those two dogs?"

"Got one of them, the beagle."

110

"What happened to the white dog?"

"It ate one of my guinea fowl."

"What did you do?"

"Didn't take a minute." Herb was a man of action when provoked. He also had the classic New England trait of describing an involved process in just a few words. I could imagine the twinkle in his eye as he issued this vague but adequate description of what had actually transpired.

"Well, I guess we'll take the beagle."

When Jean arrived home with the beagle, much of the stress of the last few months dissipated. It was the kind of relief that must be felt when the sociopath next door gets life without parole. Whether it was from his playful jumps or some other trait, the dog was given the name "Cricket", and once again, we had a pet that romped over the hills with the children.

Not that everything was perfect. Cricket, like many of his breed, was stirred to remember the dead of all wars on bright moonlit nights. His operatic howl came back from the hills in an echo – a muffled echo on warm moist summer nights and fainter but like shards of broken glass on frosty winter nights.

He also had a good eye for changes. When we moved the cows from one field to another, he barked for a couple of days until he was convinced that we knew where the cows were. This trait was helpful from time to time, for instance when a cow was not where she was supposed to be and we thanked him on those days – not on the others.

It was not our intention to ever have more than a few chickens. The bantam chickens we had bought at the East Thetford Auction were for color and with the unexpected rooster for the wake-up call in the morning. From time to time we would find and collect a few eggs. Most were left in clutches deep in the hay and were discovered only when they were really bad.

By now, Elizabeth was enrolled with some of her classmates in the local 4H program. Margaret Balch, the leader, worked with the girls to develop skills in a variety of domestic chores. Elizabeth was already accomplished in most of these areas - in others, such as embroidery and crocheting, it was clear that she was doomed to fail. When Mrs. Balch arranged for each child to get two dozen one-day old chicks, however, Elizabeth was right in the main stream.

In anticipation of the arrival I put up a few studs in the room next to the stanchions in the hay barn and nailed some recycled floorboards to them to make a wall. I ran out of boards near the top and closed the gap with chicken wire. I had an old panel door we recovered from a derelict house and mounted it in a frame with a strong spring to slam it tight. I also brought a wire to an outlet for the electrically heated water bowl, built some egg laying boxes and some roosts. We rummaged in the Squire's old farm equipment and came up with a large feeder and a gravity-operated dispenser for grit. I made a small door opening into the barnyard, one that could be closed at night.

The chicks arrived, 12 Rhode Island Reds and 12 Barred Rocks. Elizabeth followed the directions: keep warm, feed, and water (not too deep!). They thrived. Although Elizabeth had no great love for her chickens, she nevertheless spent enough time with them so that she was able to recognize each of the Rhode Island Reds and distinguish it from the others. While perfecting this ability she sat high on the top of the laying boxes so that the chickens could not peck at her legs while she scrutinized them.

As they grew, the first signs of a vicious social order began to appear. Whereas the bantams seemed to get along well, the Reds entered a competition to determine the most aggressive bully. Margaret Balch, anticipating that this behavior would emerge at about this time obtained the 4H debeaker and dropped it off at the house. I had some idea of what this might entail and asked Elizabeth whether she wanted me to give her a hand with the work. "No Dad, these are my chickens and I should do this job myself." With grim determination she headed for the barn.

A debeaker is comprised of a board that supports a short piece of nichrome wire mounted on a hinge. A foot pedal switch sends house current to the wire. When the foot pedal is depressed the nichrome turns red hot and drops down to the board. You grab the chicken in both hands and put its beak on the board, and then step on the pedal. When the red hot wire swings down, it lands on the front of the beak. As the wire touches it, the beak erupts with smoke and the smell of burnt tissue. Thereafter, the chicken has a stub of a beak resembling a hammer instead of an ice pick. Debeaking is akin to a serious finger nail trim and doesn't hurt the chicken as much as one might think, but is not a nice job. An hour later Elizabeth was back. All done.

Years later, Jean and I sat in the stands at the 1994 Winter Olympics and looked up the steep mogul course to the small figure that appeared at the very top. It was that same determined young woman. I cannot say that debeaking chickens led directly to Elizabeth's Silver Medal at Lillehammer, but determination is something that is acquired, for instance in acting as a surgical referee for fighting chickens.

For much of the year the chickens roamed the barnyard and adjacent fields foraging for worms and insects. I'm not sure this made them less aggressive toward each other but it was harder for the bully to corner and injure a weaker bird. In the depth of winter, however, they were confined to the chicken house and it was at this time that some disgusting injuries took place. During the day they whacked at each other, then during the cold nights they crowded together on the roost gaining a bit of heat from each other. The brain of a chicken doesn't cope with the conditional: "If I peck at the chicken who is keeping me warm it may die and I will be cold." I wonder if scans of human brains will eventually show that people who are inclined to attack their neighbors as sport, lack the same portion of the brain that is absent in some chickens.

Letting the chickens roam freely during the day had other benefits. Because they ate all kinds of food instead of a steady diet of laying mash, the egg yokes were a fine yellow, the result of the carotene obtained from discarded squash, pumpkins and other vegetables. The rare lobster shells, bread crumbs and other food garbage went to the chickens. What they wouldn't eat went into the wood furnace. Because chicken farming is big business and because chickens are easily susceptible to scientific study, a tremendous amount is known about the nutritional requirements and various diseases of these birds. It is not just by accident that the first vaccine effective against leukemia in animals was developed to protect chickens.

In the commercial world any deviation from a carefully calculated and prepared pelletized diet almost inevitably leads to slower growth and poorer egg production. A delay of a day or two in the time a chicken meets the minimum criteria for frying is enough to put a large chicken farm under. Chicken nutritionists would be dismayed at the way we fed our birds. We were happy to forego a few eggs so that the yokes of the rest would be deep yellow. I read a letter in the British journal, *Lancet*, suggesting that earthworms, a component of a free-range chicken diet, are high in polyunsaturated fats. I have not come across a definitive article on this but it may be true. Certainly, I have never heard of a robin suffering from coronary artery disease.

Soon we had too many eggs and Jean became the neighborhood egg lady. Selling eggs was far more pleasant than collecting them. The egg collector had to observe the hen-peck lesions, cope with a water bowl that occasionally froze despite the electric heater, and worry about the chickens' cold feet.

Then, one by one, the chickens began to disappear during the night. We were baffled. Most of the time there was a missing chicken and no trace of what had happened. Sometimes a search revealed one or two feathers on the barn floor outside of their enclosure. Cricket was not suspect. He never had any interest in the birds and he could not enter the room where the chickens were housed. We postulated various intruders but had no evidence to back up our theories.

One morning, early, I went out to feed the cows and heard an animal drop to the floor of the chicken house. I flung the door open and found a raccoon, chicken in its mouth. He had pushed the chicken wire near the top of the wall aside, grabbed his bird and was headed out. Surprised, he dropped the bird and retreated into the room where the baler was stored for the winter. I was right behind him, looking for something that could stop him but not finding it. He headed toward the dark corner where the barn met the silo.

This silo was one of the Squire's improvements to the farm - a nicely built redwood silo held together with iron hoops and supporting a galvanized metal dome. It was placed on a concrete base to which it was attached by bolts embedded in the foundation. A series of small doors is built into the redwood staves. These are covered by a vertical appendage on the side of the silo, Silage shoveled from the silo through the openings falls to the floor of the barn. It the lowest of these doors was open, providing access to a pit in the concrete foundation about four feet deep.

The raccoon went under the baler and through the door of the silo. There was for the second time that morning a soft thump as the animal landed at the bottom of the pit. About this time, the children emerged from the house to feed their horses and I shouted for Ross to bring me a flashlight. He came running with the flashlight and together we peered down the light beam into the darkness. Looking back at us, eyes brilliantly reflecting the light, was a very angry chicken thief. "Hold the light on him Ross and if he tries to jump out, slam the door on the silo. I'm going for the gun."

This photo is put here so that you could imagine the shotgun going off inside the silo. Carina, Sheik, and Stormy are left to right up front, and the hay rake and tedder are behind them after being pulled through the open barn door. Tami is walking toward the open door which may be shut before she gets there. The Case VAC is a bit of orange in the bay of the cow barn. The Ford 9N and hay wagon are out of sight in the bay behind Carina.

It took me a few minutes to find the Squire's 410 shotgun and another few minutes to find a shell. When I returned, the animal was still there, no happier than before. "I'll hold the light, Ross. You get out of the way in case there is a ricochet."

He turned away. BLAM!

It is at this point in the tale that I ask listeners whether they have ever heard a shotgun go off inside of a silo. The response is uniformly negative. I then continue, "neither had Jean or the horse that she was leading past the silo at that moment." Jean had no idea of what Ross and I had been up to. It was lucky that her arm was not pulled right out of its shoulder socket.

This event stabilized Elizabeth's chicken flock at a reduced number. There were no other mysterious disappearances. Elizabeth allowed Ross to give one to the Boy Scouts, however. Every winter the local Boy Scout troop participated in a weekend of winter games referred to as the "Klondike Derby". In preparation for this event, Ross and his troop needed practice in winter camping and I was drafted as the responsible adult to supervise this activity. The boys had built a dogsled-like superstructure upon an old toboggan, and wanted to practice with it a couple of weeks before the competition. We made ready and loaded the sled in the truck along with the tent, sleeping bags, cooking gear, and other cold weather items. About this time, Ross, who was responsible for bringing the meat for dinner, realized that he had none to bring. This is when his sister agreed to provide the hen.

After parking the truck at the end of the road to Bill Nichol's Christmas Tree farm we piled things on the sled and set off. The hen was able to look excitedly about from the top of the load – she had no inkling of the plan for the rest of the evening. The rest of her body encased in a tight cloth bag, she traveled well through the fading light on a cold January afternoon. As we trudged through the snow the boys became increasingly apprehensive about the form dinner might take. By the time that the sun had taken its plunge below the western hills, the cold was full upon us. We found a good spot for the tent and fire, and the boys, leaving the well-groomed Christmas trees, scouted firewood in the nearby woods.

Soon we had a roaring fire going and had prepared a large kettle of boiling water. "O.K." I said. I'll take care of the first part. You guys have to pluck it afterwards. Then I'll clean the carcass, and you cook it. Fair?" Because there was very little for dinner without the bird, they agreed.

I stepped into the darkness and wrung the hen's neck. We plunged the bird into the boiling water to loosen its feathers and many hands then plucked it. The hot water was the first warmth their hands had

117

felt in an hour, but it cooled fast. We put the bird in the pot a couple of times more as ice formed on the skin before we finished plucking. I removed the innards and we pushed a spit through the bird. Ideally the chicken would have been roasted slowly, a scout turning the bird with a stick as it hung by the side of the fire. Instead, we put the bird directly over the fire, the last few pinfeathers gave off little puffs of smoke as they encountered the flames. We ate the carcass the way that the Chinese eat roast pig, in layers, a little bit at a time from the outside-in as the heat slowly worked its way through the bird. No one complained and they learned their lesson: chicken does not really originate in the supermarket.

Eventually, the egg production dwindled, then stopped as the hens aged. The time had come to draw this part of the farm operation to a close. We picked a lovely late February afternoon when the temperature outdoors was the same as it was in our refrigerator, about 40 degrees. The old butternut tree near the garden had a large, low, partly horizontal limb that threatened to break our skulls when we forgot to duck while mowing the grass. We set up the camp stove on a small table near the tree, put a huge pot of water on it, and laid in a supply of hay rope. Now, that limb came in handy for hanging the birds - a piece of the hay rope to both feet. I wrung and chopped heads and hung. Jean dipped in boiling water and we both plucked. I cleaned. The job finished just as the sun went below the horizon.

A few days later we had a gala party for the Cancer Center staff. The birds had been simmering for about 36 hours on the woodstove before vegetables, seasoning and wine were added. As the crowd arrived, the house was filled with the savory smell of Coq au Vin. Dinner was a huge success, the meat delicious and tender. We never told anyone how long it took to cook. Elizabeth, to this day, remembers that the Coq au Vin smelled a lot better than the debeaking. Every remaining scrap of that 4H project was consumed before the evening was finished, the engineering of our exit from the chicken business completed, and the economics of chicken farming confirmed.

The butternut tree decorated by human "ornaments" during a visit of our friends, the Morans. The long semi horizontal lower branch of the tree was the one that was used when we left the chicken business.

Chapter 8

Sheep and Barns

Walt Record once emphasized to me the importance of selling one's best animals instead of the worst. This practice builds the reputation of the seller and ensures continued sales at good prices. Walt's skill in marketing went beyond this, however. Instead of showing his cows in freestalls or a milking parlor, he showed his cows in a barn with a pipeline setup for the milking machine. Here the cows stood on mats a few inches above the alley behind them. To an observer standing in the alley, cows in this setting appear larger than they really are. (I have noticed the same technique used at a successful farm equipment dealer. The tractors are displayed on a concrete platform that is a foot higher than the road from which the customer examines them. This method alters the appearance of the tractor so it looks like it has about 20% more horsepower.)

Although Walt acquired a good reputation for his dealings with cows, he felt no need to establish a reputation as a dealer in sheep. Walt had sheep only because they helped fill the slack during mud season, a time when the ground was unworkable and he was free between milkings. During these few spring weeks he could get the shearing done and he was available to help ewes that needed assistance during lambing. For the rest of the year, apart from controlling predators and coping with the numerous health problems that sheep experience, the flock took little of Walt's time.

Jean and I bought our sheep from Walt while we were still in our first house in Lyme. We wanted their help in clearing the brush-filled acre behind the house, and we thought that sheep would provide a pleasant and easy introduction to animal husbandry for the children. Walt was only too happy to sell us three ewes. In doing so, his goal was to rid his flock of its three worst animals, presumably with the idea that anything that we had to face after this introduction would be easy.

Now, years later, I think about how he must have chuckled as I drove away in our station wagon, carrying off three of his liabilities. The check in his pocket must have been a real comfort. But Walt was not doing this maliciously. He knew that if we were going to get involved with farm animals we should do it in way that taught us a lesson - a lesson that was obvious, inexpensive, and memorable. We, of course, had no inkling that we had purchased an important lesson in this transaction - not when they ran off during our first unloading, nor when their lambs found the road from the uterus to the great outdoors constricted by narrow bones in the pelvis, nor when the offspring, once landed, displayed a considerable lack of charm. We struggled – We attempted to inject new genes into the flock by using several different rams. We failed and failed again but, eventually we learned Walt's lesson: when purchasing animals select those with good genes.

We had already had some instruction in the hard-knock school of lambing. The Smiths started with sheep before us, and Mike and Jean's lambs started to come first. It was a cold and raw week. I got a call from Jean Smith's father. Mike and Jean were away and he was housesitting. One of the new lambs was failing. I drove over to their farm and tried to get some warm milk into this cold and weak animal, but it would not suck, even after I warmed it up. I watched as it weakened further and died. A week or two later, we moved our ewes to the field where I had propped up a large packing case as a three sided shelter. Our lambs started arriving soon afterwards.

It was more of the same. The wind whipped by, invading the corners of my crude shed. One of our lambs cooled quickly and failed to suck at the teat or at the bottle I offered when nursing failed. As I returned to the house in the windy blackness late at night, a dying lamb in my arms, I spotted an unused aquarium sitting on a shelf in the garage. Next to it was the aerator for the aquarium complete with a short length of plastic tubing to carry the air into the water. I pulled off the tube, heated some milk, sucked it up in a large syringe, passed the tube into the stomach of the feeble lamb, and rapidly infused about four ounces of warm milk. The results from this nourishment were astounding. Within the hour the lamb was a strong healthy appearing animal able to pursue its mother's teat with success. Some years later, farm equipment companies started to sell a "lamb saver". It consisted of a syringe and a length of rubber tubing.

Well, we hadn't learned much, but we had learned something. I stopped over to see how Walt handled this kind of problem. I entered the ground level of his large gambrel roofed barn and found him with his cows. He took me around the bull pen where the herd sire was housed behind stout bars of galvanized piping. The bull's muscles rippled beneath his well-groomed hide and air whistled in and out past the ring in its nose. Walt's barn was built into the hill on that side and behind the bull pen there was a narrow room between the stone foundation and the cows. The floor of this room was covered with a thick layer of hay and Walt's many sheep lounged comfortably there, some with and some without lambs. It was warm, the sheep looked at ease. I was beginning to understand another part of what Walt had intended to teach me. I just never had thought that I would need a barn built into the side of a hill with eighty Holsteins heating in it to make lambing go smoothly.

Walt Record's barn was built by his parents after they moved to Lyme in 1940. It replaced an earlier structure that the Lawsons, the previous owners, had lost to fire. The Record barn was one of the largest in town, and presented an unpainted façade to those passing along the East Thetford Road. Its long axis was oriented east and west; the sun beamed in through multiple windows on the south. The north side was dug into the hill and there stone ramps allowed hay wagons to be drawn up so that their loads could be emptied on the second floor. To feed the cattle, hay was dropped through openings to the ground level, an open floor with well-spaced posts holding up the two higher stories. In the floor a chain-driven cleaner swept gutters that ran behind the cows and conveyed the manure to the outside. There it could be loaded easily for spreading.

This type of construction was typical of the 1920's and 30's when electricity to drive gutter cleaners, and reliable tractors to haul large spreader loads of manure, were becoming available. The big barn that we acquired with the purchase of the Squire's place was quite different. In fact, from the arrangement of the timber frame, it was clear that it had been formed by placing two separate timber frame barns end to end. One was comprised of timbers that had been hand hewn and the other from timbers that had been sawed. Although the north end of our barn had some earth heaved up against the foundation, this arrangement did little to blunt the wind that whipped down the Connecticut River valley from the north. It scoured in along the west wall, raising miniature dust devils in the loose hay. There was no evidence that cattle had been housed or milked in the ground floor of this barn.

Instead, there were plenty of clues that cattle had occupied the second floor of the barn. On the north end, the floorboards were hoof-worn and covered with a powder of ancient manure. Housing animals on wood in an upper story of a barn was a practice that developed after 1850 when it was recognized that this labor saving arrangement allowed the manure produced by the animals to be dropped through hatches to a storage pit below. In this protected area rain and melting snow did not carry away the nutrients from the manure. When spread later, the nutrients were still in the manure to nourish the hay or corn crop.

Even if we had desired to update our barn so that our cows could be housed at ground level, the juxtaposition of the two barns to make one large barn would have left a profusion of supporting posts at inconvenient distances so that the arrangement would never have been satisfactory. As I thought about the differences between Walt's barn and ours, I was beginning to understand the important relationships of construction to microclimates. Bit by bit, and over a couple of hundred years, the human experience in this corner of the world taught people how to take advantage of sunshine and the inherent heat of the earth, and how to make the best of it with local building materials.

We never had any intention of trying to house our sheep or cattle in the large barn. It became the "Hay Barn". The smaller barn, which had also housed cows, we called the "Cow Barn". We tightened up the room at the east end of this barn, the whitewashed room with a small paddock to the east. Although I never got all the doors to shut tight, the room got some sun through its two south windows, and with hay stacked in the mow above as insulation, it was the warmest spot we could find. I built a couple of lambing pens into one end, rigged an outlet where a heat lamp could be connected, and arranged the fence so that the sheep could get to one side of the cows' water bowl. We borrowed a ram from Hugh O'Donnell that had a reputation for fathering easily delivered twins and we struggled onwards. Our problems continued.

Each spring, during April and May there were three important national medical meetings for people in my field. I generally attended two of them. In addition, there were frequent out of town committee meetings that also required my attendance. I made friends of the passenger agents at the Lebanon airport, and my small brown nylon carryon bag was seldom put away. In the year when my service on these various committees peaked, I had 45 trips requiring one or more nights away from home. During these trips away, especially the trips to the spring meetings, dramatic things would happen in the family sheep operation.

When I opened the kitchen door on returning from one trip the musty odor of damp hay greeted me. From the downstairs bathroom I heard bleating, and the children were bouncing with joy. "Come look!" There were three lambs hopping around in the downstairs bathtub. The ewe had died.

At other times, the homecoming was more traumatic. I returned from one trip to find Jean subdued. In the middle of the night she had called Walter Record to assist her with the obstetrical problem known as a footling. The lamb lay crossways in the uterus with a forefoot presenting through the birth canal. Despite Walter's strength, it was impossible to push the lamb backwards to a head-first position. Together they decapitated the lamb and delivered it in two pieces, thereby saving the life of the ewe. This is not the usual kind of experience for women who are raised in the New Jersey suburbs and who major in political science at Smith. My admiration for her ability to cope was cloaked by my guilt at being absent.

So, we got rid of the sheep. Not before our friends, the Penfields, stopped over one day. While we chatted, their dog hopped from their car and ambled around the corner of the barn where he was found a few minutes later devouring a ewe. Friendships that can survive this experience are solid. We blamed the dog, not them.

One midnight a year or two after we had left the sheep business, Jean and I were sleeping comfortably when Jeanie entered our bedroom. She had just arrived home from a baby-sitting job. "Mom, Dad, Walter's barn is on fire!" Lights went on, clothes flew about. In a minute the five of us were in the car and on our way. It was pretty well over by the time we arrived. Hoses snaked across the East Thetford Road, a huge pile of glowing embers sent sparks skyward when disrupted by streams of water. There was nothing left but a few blackened posts and collapsed beams – those and the terrible odor of burned, and then wetted hay mixed with the sulfurous smell of burned hide and hair. I found Walter. He looked in shock, a wet, grimy tee shirt clinging to his huge torso. I asked what we could do to help, suggesting that he might need to drive some remaining stock over the hill to our pastures. In sad tones he said there was no need; all but one cow was gone – those proud cows that had stood so tall in that grand barn, the bull roasted in its galvanized pipe pen.

The next day as I came home from work, a big excavator was already there burying the dead animals, knocking down the charred remains of the barn. There was a huge hole in the view to the north where the barn had stood – a big chunk of Lyme was missing. Walt's hobbies, the exotic birds, chickens that laid green eggs, peacocks, birds that he proudly showed the children in grade school were still there, but his livelihood was gone.

Ruth Malmstrom, a neighbor of Walt's at the time, displayed the kind of conviction that has allowed generations of her Norwegian ancestors to survive calamities. Not waiting for others to act, she announced that enough money would be raised so that Walt could rebuild. Within days she had roped Jean into this project, and together they assembled a team that geared up for auctions, concerts, bake sales, and other events. The goal: $50,000. Shine King, in turn, got the word out to nearby farmers and they began to contribute animals.

That is how we got back into the sheep business. Walt was given two good Romney sheep. One, the ram, took advantage of Walt's turned head to administer a terrific jolt to his backside. He went back to the donor. The other was a very sweet ewe.

"Jean, this is Walter. You know that ewe that I was given?"

"The one that was the woman's pet?"

126

"Yes, that's it. It has eaten my vegetables and the rose bushes. Now it is working on Ada's flowers. It won't stay in a fence and follows me all over. It's driving me crazy!"

"Give it back to the woman who gave it to you."

"I called her and she won't take it back. Said that's why she gave it to me! You have good fences. I'm giving it to Elizabeth. Bringing it over now."

It was true, our fences were better than Walter's. Also our flowers and vegetables were no match for his, so the inducement to delinquency was less. The ewe came to stay, and at last we learned the lesson Walt had been trying to teach us with the first lot we had bought from him. This ewe had good genetics – it makes a difference. Our reentry was easy. Jeanie named her Jessyma. Her lambs and her granddaughters were strong, they didn't get stuck during birth, ewe's milk came easily, fleeces won prizes at the New Hampshire Wool Fair. They were better to look at, and they may even have been more intelligent, though using that term in connection with a sheep is an insult to most other animals.

Now if we had devoted 25 years to the selective breeding starting with our first three miserable ewes we might have eventually matched the quality of the gift ewe. We would have had first to know the qualities we should select for in the offspring, We would have needed the knowledge to select an appropriate ram whose genes could take us on a path to overcome the particular deficiency we were addressing, and the patience to persist in the march to our goal. In addition it would have been necessary to exclude compassion as a selection criteria and to use only objectively measured qualities for this purpose. The fact that a child enjoyed cuddling a newborn lamb would not have been relevant to the selection game. The process is ruthless but must go forward if one is to be a successful breeder. It is also unnecessary, since others have already done it for us. We just need to appreciate what generations of prior breeders have done and reward current breeders for what they have accomplished.

In breeding the flock carrying some of those Jessyma genes, we tried to avoid introducing undesirable traits. There are many ways in which things can go wrong, but thinking back to those first three sheep that we unloaded from the Travelall so many years ago, we were way ahead.

Nevertheless, sheep do get sick, and they die in all kinds of ways, including some that defy logic. We kept in mind the price paid for sheep at the East Thetford Commission Sales and balanced this against the cost of visits by the vet. We became hardened to the realities of producing good fleeces, decent chops, and ewes that had easy deliveries. We found humor in our misadventures, laughed at our significant failures, and have moved, I fear, in a direction that upsets those who would forgo eating lamb chops from a culled lamb. To those who criticize this "brutal" approach to animal husbandry suggesting a replacement vegetarian diet instead, I point out the massive sacrifice of inferior soy beans that was required before those yielding decent tofu could be selected.

You will note the euphemisms in this discussion. Instead of saying that we took lambs to the *slaughterhouse* in Sharon, Vermont we used the euphemism "Sharon Beef", and instead of saying that we sent failed ewes to the auction we employed the word, "cull". When talking with my farmer friends about culled ewes and rams I would use the word they used, "ship" as in "I'm going to ship the ewe that had mastitis". Putting things bluntly is not a characteristic of our time.

David Hinman, who did our shearing, often takes sheep to a livestock auction. Although called the "Northampton Auction", it takes place in the small town of Whately, just north of Northampton. He was given a few miserable pet sheep by a kindly woman who did not have the courage to do what had to be done. She felt David could find them a suitable home. He reassured her explaining that, "I will treat them as if they were my own." And since if these wobbly sheep were his own he would immediately take them to the auction, he did so. He later told the woman that he traded them at the "Whately Fair". Our language and society have come a long way. A livestock auction may represent a far more lighthearted event than a county fair infiltrated by scammers and sometimes violent roustabouts – why not call it a fair? The donor, I'm sure, imagined a maypole on the greensward and dancing to the song of a lute. Peace of mind has great value.

While we are engaged in etymology, let us clear up the confusion about "lamb," especially "spring lamb" and its relation to mutton, a meat that some regard as inedible. In meat industry parlance the definition of lamb vs. mutton depends upon a singular anatomic difference between immature and mature sheep, the closure of the epiphysis in the leg. The epiphysis is the site at which bone growth takes place, a soft cartilaginous place near the end of the long bones. To tell the difference between lamb and mutton, the butcher simply bends the leg until it breaks. If it breaks at the epiphysis it is lamb (the bone is not yet done growing), if the epiphysis has closed the bone is stronger and the joint breaks. Now it is mutton. Lambs may be born in the spring, and many are. However, some breeds have lambs at other times. The definition of lamb is unrelated to season of birth, season of slaughter, religious holidays, or weight of the carcass. It is orthopedic.

When we first started with sheep, Mike Smith and I sheared a few with Mike's dog clippers. We shot a couple of weekends and our two backs. When we finished the sheep looked like they had been in a fight with an electric hedge trimmer. Since then, professional shearers have done the job. It is done in early spring before the lambs were due. At the end of winter, tags of manure soaked wool are hanging down in the groin presenting a false teat to confuse the lamb. The lamb doesn't have energy to waste as it searches to find the real thing. Shearing takes care of this.

To start the process, David hangs the electric motor on a rafter and uncoils the drive cable. His hands, softened by pounds of lanolin from the fleeces, grasp the clipper as he takes long strokes, while peeling back the clean yellow wool that falls in a carpet at his feet. His fist placed at strategic locations of muscles or joints and immobilizes the portions to be sheared. The sheep, now almost hypnotized, becomes putty in the shearer's hands. The pink skin emerges from under the heavy coat of winter, and as the last of the wool falls, the ewe sprang forth from David's grip, looking like an entirely different animal.

Although the shearing, hoof trimming, and worming takes only a few minutes per animal, there is plenty of time for talk, and there is no better subject for a shearer to talk about than sheep, unless it is about the people who own them.

David described a call he received from a woman who asked him if he could come by and shear a few sheep so that her husband could learn how to do it. David agreed to do this and a few days later stopped by the place. The couple, who knew nothing about sheep, had jumped into the business by acquiring a large number of animals. He sheared a few sheep while the husband looked on and then departed, wondering just how the neophytes were going to cope. A few days later he got a second call from the woman. Her husband had been "having some trouble with the shearing" and could he come back and do the rest of the sheep?

He returned to find a couple of sheep covered with large loose tags of wool and deep lacerations. Bending to his task he proceeded to shear the rest of the flock. As the fleeces peeled off the sheep to the rhythm of an expert shearer the husband said, "Oh, that is the secret. I know what my problem was. Your shears are much sharper than mine."

David nodded in the direction of the two miserable practice ewes, "If your shears were any sharper you would have killed those two."

As we sat having hot drinks one afternoon after the shearing, David mentioned that someone ought to do a story about all the ways sheep have of dying.

I went to my desk and came back with the following:

Sheep Die

There are really only two things sheep do really well: they make wool and they die. While the wool-making goes on almost imperceptibly, this is usually not the case with the dying, although on occasion they may die imperceptibly, too.

Usually it is more dramatic. One of our orphaned lambs, (the mother had died) was growing strong at the rubber teat on the bottle of milk replacer, crowding in amongst the other orphaned lambs to get its fair share. One day it went on sucking long after the bottle was empty until I found it, eyes glazed with satisfaction but in belly-swollen-agony, as its distended abdomen pressed on the lungs excluding air. I ran to the house for a knife while at the same time I tried to remember which side of the belly to cut to decompress the rumen. In seconds I was back, but too late, the lamb was dead.

One of our ewes had four lambs, one after another - pop, pop, pop, pop. Then surveying the scene and in awe of the task ahead she lay down plop! - and died. I have learned that you should never allow a sheep that has an excuse for dying the opportunity to lie down. If you can keep it on its feet it will usually not die, but I do believe it is possible for a sheep to die standing up.

Often they die on their back snuggled into a warm corner of the barn. They lie there with a heavy load of lambs and chew their cud, until for some reason they decide to roll over against rather than away from the barn. Now stuck on their backs, the vent from the rumen is closed by the anatomy of their predicament. They fill with gas, until some hours later they are found stiff with rigor mortis, belly huge, legs extended.

At times they die in even more dramatic fashion. One of our lambs was identifiable by its crooked face, the result of accidental inbreeding. It prospered despite its genetic handicap and by August was a wonderful specimen as long as you did not look at her straight on. The White Gate Pasture where we had the flock was grazed down to nothing so Jean opened the fence into the area along the brook, where amongst the alders there was a crop of fine grass. For several days the sheep were here, eating the rich grass and lying in the cool shade of the ferns, twitching their ears at the infrequent late summer flies.

One evening when we had something important to do (they always die when you have something important to do) we found that Crooked Face was missing when we put the sheep in the barn for the night. As the sun was setting we began a systematic search of the several acres along the brook and eventually found the scene of a tremendous struggle. Bits of wool lay about. A huge log, long bedded to the soil to rot was ripped loose. Ferns for 10 feet around were crushed and torn up, and later after much more searching, crooked face - just the face and head – was found, looking at us from under some brush. Silly grin! It inspired me to murder, but the bear that had done the deed was long gone. By the time we could get back to clean up the wreckage, there was nothing left. The coyotes had come through.

The cleanup is seldom as easy, however. Twice we have been fortunate to have large brush piles to burn and it has been necessary only to lift the carcass to the top of the pile, call the fire warden and Hanover Dispatch to tell them we are burning, and set the fire. A good fat sheep will actually help a pile of marginally combustible wood get going. Getting the sheep to the top of the pile is not easy, however. One of my patients claims that a sheep doubles in weight the instant it dies, and this is approximately true. Until rigor mortis sets in, the task is nearly impossible. The head hangs down, dragging on the ground, preventing any effective heave-ho. Gas and feces pour out at the other end. It is a thoroughly disgusting job, even if one has a ready brush pile.

Of course, you can always bury the sheep, if the ground is not frozen. Folding a limp sheep into a hole deep enough to discourage digging animals that may pass shoots half a day, but it is good for the soil. If you don't have the time you can put the sheep in the bucket of a tractor and head for the most distant part of the farm, hoping that the bait will not draw vermin. I have been delighted to see turkey vultures appear in Lyme. Never saw them before. (Must be more people going into the sheep business.) Maybe they will get the sheep before the coyotes.

But when there isn't a brush pile, when the ground is frozen and when the snow is too deep to head for a distant part of the farm with the bucket loader, that is when the trouble begins. Hauling a sheep on a toboggan is not satisfactory. A dead sheep has more ways of falling off a toboggan than a live person has. A day like this is a bad day for a sheep to die, but, as I said, sheep hardly ever die on a good day.

Jean's cousin lives in a respectable community of suburban estates and obtained a few sheep so as to complement the motif created by garage cupolas and sports utility vehicles. The ground in New England suburbs freezes almost as hard as it does on the real farms, so when one of his sheep died in the middle of winter, he found his shovel failing to dent the prospective gravesite. He called his vet in hopes of discovering an animal crematorium.

"Ground won't be frozen under your manure pile," said the vet.

The cousin hung up, returned outside, and contemplated the large pile of steaming manure while doing computations concerning the time and effort required to uncover the required gravesite.

Here the story gets a bit confusing. Some say he moved the manure and dug the grave. Others say that he was suddenly inspired by the lifelike pose the frozen carcass had struck, and under the cover of darkness hauled it to the lawn of a neighbor where he propped it up amidst the group of plastic animal replicas that decorated the neighbor's estate.

So sheep die easily. That is a downside. But there are upsides that make up for the dying. Roast lamb more than makes up for all the middle-of-the-night expeditions to the barn to check on ewes during lambing season. In addition, I believe that as families in well-off communities fork their lamb chops up from the platter, they should understand the value of lamb in the diet of those in many poor world communities. In these, celebratory lamb or goat feasts may constitute the rare, perhaps the only, decent meal of nutritionally complete protein ingested by an impoverished family during the year.

Also we should not forget that among us there are those who cherish the clean yellow wool that is in the arm-load of fleece that their shearer has just picked up from the shearing floor and handed to them. Jean McIntyre was such a person. For her, that fleece was more than an article of commerce. It was access to a traditional occupation worthy of celebration. When washed, carded and spun, it could be made into a garment that often far outlived the sheep that produced the wool. She was an artist who selected the best wool from the flock for her projects. Her eyes were on the sheep within the flock that carried the fleece that she knew would be ideal for the project that was in her mind.

She would put the fleece into a large bag made from a worn out sheet and label it with the name of the sheep and the date. It was then hung to swing in the heat of the barn loft to dry. That way it wouldn't go moldy. Later, she took the bag down, spread out the fleece and cleaned it of any matted tags or hay. While doing so, she looked at the wool. She knew by now how this wool would spin and the kind of garment it would be used for. The bundled fleece then went into the washing machine that was reserved for this purpose. (The drain, connected to a pipe that went into the sheep yard instead of the septic tank. The wash water is full of things that help the grass grow.) After the fleece was dried, she carded it to align the fibers and pulled it into the loose rope-like "roving" to be spun.

Later, she would join a group of her friends all of whom spin wool into yarn. They gathered, each bringing their spinning wheel and their own roving, and while yarn was spun into being, offered advice, critique and support to those new to the game. They bonded. It was a party of learning and sharing knowledge that was good for the soul.

A watercolor by Anne Mellor shows the spinners and their several types of wheels. This craft produced wonderful friendships.

While they spun, they could converse, and the participants, after a few such gatherings, became a resource when a person in the group needed help or support. This gathering around the spinning of wool was much more than a school for advancing the art of making fabric, it was part of a happy life. The watercolorist, Anne Mellor, joined the local spinners and recorded the gathering, The raw material, wool, went to fairs and got ribbons, and when the garments came to be displayed, we were witnesses to an art.

Wool production in the U.S.A. peaked during the years during and following the Napoleonic Wars. Sales of wool had allowed the original Riverbridge farmhouse to be enlarged into an early19th century luxury home. If one had to spend the year out in the weather while fighting a war, wool clothing was a necessity for survival. Now synthetic yarns from petroleum and the like are wool substitutes. While they offer some advantage over wool, many adventurers still select wool for tough situations. It certainly feels good next to my skin in bad weather, and the sheep who use it to sleep outdoors in the snow, seem to like it too.

Chapter 9

Farm Kids

Clodhoppers is what we called them in 1942, the ankle-high work shoes that were worn by my schoolmates who had transferred from rural school districts into our city school. Our shoes were low. Clodhoppers left black skid marks when they scuffed the floor. My parents wouldn't let clodhoppers in the house.

Ill at ease, distracted by aspects of city life they found challenging, farm kids stood out from the rest of us. On the way home from school one of them walked several steps into a freshly poured cement sidewalk before realizing that something wasn't right. We laughed and referred to him afterwards as "the cement walker". (Maybe there were some advantages to those high shoes, after all.)

Drawn by wartime employment at the bomber plant or other war industries, their parents had left the farm and had settled in Omaha. We offered these children little comfort, until, like us, they wore low shoes and had melded with our culture. The fit was seldom perfect, however. Often physically stronger, more independent, and with gaps in their cultural underpinnings we found amusing, detection of their rural background remained possible over the years to those who cared to look. For instance, if you drive enough in the Midwest you will sooner or later find a driver there who pulls the vehicle far to the right before beginning a left-turn. You learn to turn that way when you have spent lots of time on a tractor pulling wide and long farm equipment.

*****

These memories floated upwards in my mind as Jean and I sat under the hanging lamp and considered purchasing the farm. If we bought the place, our children would grow up to be different from children raised in town. Despite the fact I was not a real farmer, and Jean was not a real farm wife, the day-to-day life the children experienced would flow from the farm and its work. More work and less play. Fewer playmates. More exposure to adults. Less time with the arts. In return, they would be immersed in nature – closer to nature's beauty as well as its brutality.

There were other issues, too. Adolescence was on the way. The concept of romantic love is not burnished by the observation of barnyard animal breeding. And table manners - all of us eat faster than we should.

We had already denied them television, believing that this media provided an unsuitable framework for the socialization of American youth. Our children were left out of all discussions that began, "Did you see...........last night?" And because of this they were already viewed by their schoolmates as "different". Instead, our children read, devouring the usual children's classics as well as books by Stephen Meader, a favorite author of my childhood. Meader often based his stories in rural New England and used the rich history of the region to educate as well as entertain. His heroes overcame adversity and achieved success by dint of hard work and good sense.

Thinking back on the decision to purchase the farm, we really didn't worry about the "farm kid" issue. We simply recognized it and were willing to have our children regarded as "outsiders" if and when the time came for them to enter a more urban and urbane life. We knew that they would have a happy childhood on the farm.

The fall colors as seen from our bedroom window. After the last hay was taken from the River Meadow in September, we could put the cows there to finish up around the edges.

After our move, the children continued to read, and there was lots of play as well. On the floor of Elizabeth's new room, illuminated by the bright sunshine coming in from the south windows, the children constructed barns and fences out of their blocks, filled them with plastic animals, and invented stories that they acted out with the livestock. When the physical limitations of their blocks and animals prevented full story development, they went to paper designs, fanciful diagrams of intricate barns filled with imaginary animals, each one given a name. As might be expected, these games could not go on forever and sometimes ended in disagreements amongst the participants, slaughter of the animals, destruction of the edifice, and flying rubble. The doors in Elizabeth's room soon were pitted from encounters with flung debris. When Jeanie fought Ross or vice versa, we joked that Elizabeth's room, located between the two, became the local version of Alsace-Lorraine, overrun by armies to either side. Despite the evidence still visible in the wood-work to this day, disputes were not frequent, and there were long interludes of peace.

In the deep snow alongside the path to the barnyard, the children carved out make-believe car interiors and they drove these immovable facsimiles to the sound of spinning wheels and roaring engines. (Jeanie, older and more demure, dropped out of the racing portion of this game.) When I went up the roof to shovel off a particularly heavy snow pack, the children climbed onto the roof of the mudroom and rode a snow chute, rocketing into the big pile of snow against the house.

Sliding off the mudroom roof.

We ran some old wooden downhill skis through the table saw, cutting off the metal edges and thinning them down into "cross country skis". On these we ascended to the High Pasture and flopped our way back down the narrow woods road leaving deep "bathtubs" at each sharp corner. One of the benefits of having the Dartmouth Skiway in Lyme was the free pass it gave to all the school children in town. This removed the cost of lift tickets from consideration as children made the decision on what type of skiing they chose on a given day.

The arrival of snow meant visits by our many friends from the city. The Dartmouth Skiway broadcast Swiss and Austrian music from the lift towers and our children joined the four children of our frequent visitors, the Morans, as they attempted to ski as well as children in the countries from which the music came. After an exhausting day on the hill, the children gathered around the ancient pine table in the kitchen for dinner while the adults withdrew to the dining room. Peels of laughter coming from behind the closed door sometimes caused the adults to audit what was transpiring. On one such occasion, we found all seven children with huge orange colored grins, quarter slices of orange peel behind their lips and covering their teeth. On another, the ceiling had been peppered with peas – the old pea-in-the-napkin snap.

A photo shows the seven children climbing the limbs of the Butternut tree. Another shows the children draped over the various body panels of the Squire's pickup truck. All of them good students, the vacant expressions on their faces in the truck photo suggest low test scores.

When these days wound down, the unheated attic over the kitchen was filled with sleeping children, crowded as it was when this portion of the house provided tavern beds for those on the road to the north. In the morning, the kids would emerge from the attic, trading the ice-crazed windows for warmth, and prodded toward breakfast by the sounds of the massed pipers of the Coldstream Guards pouring forth from the Sears Roebuck "Silvertone" record player.

Elizabeth, the youngest, often had to hurry to catch up with her older brother and sister. We finally gave this victim of repeated hand-me-downs a new pair of skis, which were stolen on the first day that she had them. Crushed, she returned home from the Skiway in tears. A second new pair did not seem like a good idea despite her unhappiness. The Moran boys were with us as we rummaged through the barn looking for a solution to the problem. It came in the form of a battered pair of orange Siderals, a ski of unquestioned excellence according to the older Morans, despite the fact that the butt of one had delaminated. With this recommendation stanching the flow of tears, we took the skis to the workshop where two copper rivets were let into the plastic bottom and peened on the top, binding the flapping butt together. In moments we had made the young girl happy, and her skiing skills prospered despite the dilapidated condition of her equipment.

Jean bought a fine upright piano from a woman who was moving from the community and soon we could listen to Jeanie as she practiced her Sonatina. (Family priorities had put the new piano ahead of replacing our "Silvertone".)

We clamped and glued three pine planks into a two-inch thick panel and Jeanie traced an outline of the river and a bridge on it. Carefully she then penciled the letters spelling "River Bridge Farm," the long arm of the "F" being part of the river-bank. When her lines were cut deep with a router, then painted, and the boards varnished we could hang a wonderful farm sign, probably the only one in the state coming from the pencil of a 12 year old.

Jeanie laid out the drawing on pine, and a neighbor routed the letters and picture.

The sewing machine was usually out and there was plenty of fabric about as Jean made and repaired much of our clothing. Soon Jeanie was sewing as well, doing more complex projects month by month. Eventually her babysitting jobs gave way to sewing prototypes for a local children's clothing firm. Despite the attractions of fabric between female fingers, and the lure of calligraphy, Jeanie continued to lift hay bales, train her two ponies and a horse, and, one summer, raised milk-fed veal calves for the market.

Ross was a careful observer each time Joe Bill came to the farm to repair our equipment. Joe had a van full of tools and parts, including a welder, and was willing to tackle just about any job out in the open using the tools at hand. From Joe we learned that many jobs that looked impossible could be successfully done, but only if one was creative. For instance, the frame of the manure spreader had broken apart, the result of overloading. The two frame ends were an inch apart, held from complete disruption only by the wood bottom of the spreader and the apron chain. I figured it was done for, or as the Tullars' hired man, Dell, said about equipment in this condition, "it had bit the radish". At the dinner table Ross excitedly explained how Joe had brought the two ends of frame together for welding by rigging a come-a-long between the butternut tree and one end of the spreader while using the tractor to restrain the other end. We still use the spreader today.

With these observations as background, Ross combined the engine and transmission of a cast off garden tiller with the front end of a worn out motor scooter in order to make a three wheeled ATV. Adapting automobile wheels and snow tires to the axles on this rig provided ground clearance and improved traction. Despite seeing him chug around the farm, I was not prepared for his bushwhack trip up Smarts Mountain, the highest peak in town. With pride he showed us the photos of his rig taken at the summit to substantiate his claim to the mountain top.

143

Soon, the Tullars sought out Ross as a part-time employee, arranging for him to attend the mandatory farm safety and tractor-driving course required of all under-aged farm employees. As a result, our conversations were enriched by the pungent phrases used by some of the other farm hands, and we learned something about the business of running a large dairy operation.

With Jean as the darkroom instructor and several ancient cameras available to the children, their education in photography began. They learned that cats, especially kittens, made excellent subjects. I frequently arrived home to find the print washer gently agitating new prints in the downstairs bathtub while the photographer waited eagerly to show me the photo of the current dynasty of felines. One of the kittens later lost her hearing and was unable to protect herself from unseen threats. While asleep beneath the wheels of the truck and not awakened by the starting motor as I backed out of the garage, her demise was tempered by the photographs young Ross had taken of her. As we discussed this accident, the many risks faced daily by their deaf grandmother and other deaf people came into focus for them.

Tom Moran, who spent a summer with us, took up darkroom work as a sideline to constructing a bike race track around the vegetable garden, thereby predicting his later career as professional photographer of bike racing events. He was also drafted as a pitcher on the Lyme baseball team, where his scorching fastball defeated the otherwise much stronger Hanover team and left them speaking of the "ringer" that Lyme had brought in. Although Elizabeth participated in Little League, our crew was relatively inept at baseball. As was later pointed out by Elizabeth, we could only show our athletic talent if the sport involved the feet instead of the hands.

When it came time for high school the children often had to arrange for their own transportation. Lyme did not operate a bus route to Hanover High. In the morning I frequently could drop the children on my way to work at a spot where they could walk to the school. It was the ride home that was difficult to arrange. They tracked down rides somehow and made it work. This put the children in close proximity with a number of adults. The drivers came from various backgrounds and offered a gamut of driving abilities. A critical assessment of the driving skills of various ride-givers often boiled to the surface of the dinner table discussion. Our anxieties about safety were offset by the fact that the children were forced on a daily basis to deal with adults of diverse backgrounds. The ease with which our offspring deal with their seniors today, I believe, had its origin in finding the solution to and surviving the rigors of these numerous transportation dilemmas.

After watching the interest level of Jeanie and Ross plateau in the seventh and eighth grade when the Lyme curriculum ceased to challenge them, we insisted that Elizabeth transfer to the Hanover middle school after she had completed the sixth grade. What followed was the rebellion that one might expect from a third child – rebellion and tremendous unhappiness. The intellectual and social challenge of the larger school, however, won out in the end. When she moved on to high school a couple of years later Elizabeth fit into the social structure more easily than our other two children, being elected to the student council and surrounding herself with a larger circle of good friends than did her sister and brother.

As the children grew into young adults Jean and I developed the reputation as the "meanest parents in Lyme." Early on, we established a "24 hour rule", that is we required 24 hour advance notice for attendance at all significant social events, such as a sleep over, before we would give a child permission to attend. Since many teen-age social events are quite spontaneous, we issued many such denials. Those that provided 24 hour notice could be checked out if necessary. Once enforced, this rule was accepted with only modest complaints. From time to time our position allowed the children an exit from what they perceived as a dubious event without loss of face with their peers.

This "meanness" was exhibited in other ways, as well. Because I was concerned about the risks (physical as well as lifestyle) facing downhill ski racing stars, I promulgated another rule: No skiing in contests where time determined the winner. This rule was of no concern to Jeanie who was content to enjoy recreational skiing. It may have discouraged Ross from pursuing the reckless life that some of his peers launched from a background of full-out descents. (He participated in such descents, they just were not timed.) Elizabeth took another tack, requesting permission to engage in freestyle mogul competition. In this event only 25% of the score is determined by time. Her negotiation skills were rewarded, but with the stipulation that she not take part in contests that involved inverted aerials.

Give-and-take negotiations also culminated in a small log cabin built in the woods by Ross and his friend Steve Wurster. One summer night while driving along River Road I heard a boisterous racket made by Ross and Steve filtering down through the woods from their cabin. It was high-spirited-jinks of some sort and I thought about running up there to find out what was going on. I'm glad now that I did not charge into the woods to turn whatever it was off. It was hard not to do so at the time. They turned it off themselves and no harm was done.

Ross and Steve Wurster built their cabin using hemlock logs left over from a neighbor's timber harvest. It still stands today,

Not to be outdone, Elizabeth constructed her own log cabin up near the High Pasture. Her cabin was farther from the road than the cabin that Ross built, and her friends were more restrained. Instead, we wondered whether the group of girls sleeping overnight

146

in the remote cabin was safe from predatory harm. The knowledge that our German Shepherd, Smokey, was sleeping on the floor just inside the door provided us with some comfort.

Thus, the "breaking away" process for our "farm children" was different than it would have been if we had lived in town. More difficult perhaps because of restraints imposed by our parental rules, easier because of the outlets that the farm provided. Throughout these years, the children remained entirely committed to our collective enterprise. They never asked to be excused from their share of the duties. That simple talk with the children about how the farm would change our lives that we had before we purchased the place had set the direction of our shared enterprise. Over the years they took on progressively more responsibility and put more muscle into work that made the farm a success for them and for the parents.

Jean and I could look at 50 tons of hay in the barn at the end of summer as more than a commodity. It was there because, though the children were preparing to leave the nest, their commitment to family and the farm remained solid.

Chapter 10

Tractors

During our first winter on the farm, I began to plan for the work I could see coming the next summer. The steps required to turn grass in the field into hay in the barn were conceptually easy to understand. I received abundant notice along the way, however, that the devil was in the details.

I had by now driven the Ford 9N tractor I bought from the Squire enough to know some of its strengths and weaknesses. It was small, close to the ground, and was the first American tractor to have a three-point hitch. This hitch, an invention of Harry Ferguson, an Irish engineer, bound the implement to the tractor reducing the risk of backward flip-overs, and allowed the operator to pick up an implement and drive off with it. This tractor was an instant success. It was designed to replace the two-horse or mule teams prevalent on U.S. farms. One of the Ford publications showed a farmer asking the bank manager for the loan to buy a new 9N. The farmer probably was basing his argument on the efficiencies in the switch to mechanical instead animal power. That photo was from the last years of the Great Depression and both the farmer and the banker appear malnourished.

The measure of safety delivered by the three-point hitch was important to me. I had had a childhood acquaintance killed in a tractor roll-over, and another one, (an M.D in the Midwest) was killed during the years described here. And still another friend only two months ago. It is easy to ask a tractor to do something that can kill you. You must learn not to ask. The course that young Ross had to take before he could work for the Tullar farm was devoted largely to instructions that would make operating a tractor safer.

In contrast to Ferguson's engineering genius, Ford's 4 cylinder engine on the 9N was based on outmoded designs and its six volt system made it hard to start on cold days. In warmer weather it often seemed not eager to work very hard. More efficient overhead valve engines by now were common in competing gasoline tractors and I can imagine Ferguson's impatience with Henry Ford's old engineering. The Squire had put a large quality photo print (like the one movie stars send out to fans) of Henry in the Ford booklet about the 9N tractor. He is sitting on a 9N seat with a hand on dash, and his famous straw hat astride his thigh. Perhaps the Squire and Henry were kindred spirits.

At a farm auction in Randolph, Vermont I had already purchased a modern hay rake with a three-point hitch. A week or two later, Dick Jenks and I went to pick it up. We bumped over the frost heaves as we crossed the hills on blue highways to get there with his dump truck. The farmer placed the rake in the truck body with his loader and we came back via the Interstate talking over the condition of the world as we cruised along. He refused payment for anything but the gas.

Only when I got the rake home did I realize that this three-point hitch hay rake was a bad idea. Mounted under the belly of the 9N was a Ford mowing machine designed for heavy-duty mowing, especially alongside roadsides and highways. It was driven from the power take off by a shaft that ran under the rear axle. Because the mower was connected to the three-point hitch, I had to remove the mowing machine if I were going to pull the three-point hitch rake with the 9N. I should have bought a rake that attached to the tractor draw bar instead.

Well, maybe, I could mow, remove the mower, and then rake. I would still need a larger tractor and a baler. After raking I could reinstall the mower for the next piece of work. So I pulled the 9N over to the door of the horse barn on a sunny afternoon in the early spring to practice removing the mower.

I spread out my wrenches along with the Ford mower instruction manual. As I proceeded, I found pencil notes in the manual from the Squire. He or his hired man had done the same thing, checking off each step in the process. An hour later, I still had a way to go. Even with practice, it was going to take almost 2 hours to get the mower off, and as long to put it back on. I wiped my greasy hands, thought over how infuriating it would be to struggle with the mower while rain threatened hay that was ready, and decided that to make hay I needed to solve the rake problem and to buy a second tractor. Slowly, I began to reassemble the parts that I had removed.

When I attached the drive pulley to the power takeoff shaft, I was close to having everything back together again. Absent mindedly, I tightened the bolts that held it to the shaft. Suddenly, there was a sharp crack. In disbelief, I looked at the fractured cast iron pulley. I had tightened the bolts too much. Then, I remembered something. In the back of the horse barn, covered with some of the Squire's motor oil barn treatment, I found it. A power takeoff pulley cracked just like the one I was looking at on the tractor. The Squire had done the same thing! Damn, how could I be so stupid?

Lockwood Reed in Hartford, Vermont sold Ford tractor parts as well as a variety of other farm machinery. The business, originally housed in a barn attached to a large frame house; had just moved into a new garage and repair building on the same site. I found Lockwood on the phone, standing alongside his cluttered desk. As I was to learn he was frequently on the phone.

I held up the fractured pulley. Without pausing his phone conversation he nodded his head, the gesture somehow conveying that he not only knew the part but also that he could be helpful. Finally, the phone conversation ended. Without a word, he beckoned as he made a beeline to the ell of the old house. We went in the door, up a dark stairway and down the hall to a small room littered with parts of machinery. He rummaged amongst pieces on the floor for a moment and then, in triumph, came up with a new pulley.

Back in the sun, he handed me the pulley. "Tightened the bolts too much, didn't you?" The accusation was made in the tone of a teacher, not a disciplinarian."

"Yes, I wasn't paying attention."

"Lots do it."

"Well, I won't be doing it again."

I then explained my dilemma with the three-point hitch rake and the need to remove the mower if I were to use the 9N to rake.

"You will never be able to make hay with the 9N. You need another tractor. Just sold a Ford Jubilee that would have been perfect."

A week or two later I followed up on an advertisement in our local paper by a farm equipment company in Passumpsic, Vermont, an hour up the Connecticut River. The owner, Duncan McLaren, was expanding his line of Case tractors and other equipment. I felt that with my Scottish ancestry maybe we would get along, so I gave him a call. Duncan answered. "I've got a Case VAC in the shop. We are just going over it, and I think it will be just right for you. We'll let it go for $750. I drove up Interstate 91 until it ended in a jumble of construction machinery and a view of a cleared right of way through the brush and rock outcroppings. There, I moved onto U.S. Route 5 and followed it to McLarens. In the well-lighted repair shop there were several large Case tractors all painted in Case cream color and in various stages of disassembly. Dwarfed by the others was a small Case VAC, painted orange, with a small bucket loader mounted to its front. The "overhaul" had consisted mostly of a new muffler, a steam cleaning and some new bright orange paint. The tires were good; the seat well sprung. The Case hitch, resembled a three-point hitch enough so that the hay rake, plow and other three point implements could be mounted on it. The hood, which had taken a severe blow at sometime in the past, had a sharp crease on the left side. The engine sounded good. Without trying to negotiate, I bought it.

I was mowing the oats River Meadow on a lovely day and decided to stop and take a photo. I found Elizabeth and posed her on the Case VAC.

Now, I needed a baler. I spotted a flyer that announced a farm auction in which a baler was to be sold. Although the Grays handled most farm auctions in the area, this one was in the hands of an outsider. Perhaps it wouldn't be as heavily attended as a Gray's auction and there would be some good buys. It was to be held on a weekday morning in Hartford, Vermont. After several years of spending both Saturday and Sunday mornings in my office, the laboratory, or making rounds in the hospital, I decided that I would take that part of the day off. I rose early, stopped at the office and made rounds before I drove to the auction. I changed clothes in the truck, and stepped out into the assembled crowd. I probably looked like a guilty professor stealing some time, but I didn't feel like one.

As the small items were sold, I began to understand the structure of the sale. It gradually became clear who was who, and I was able to identify the hired hand who worked on the farm. Feeling that I would be able to learn more from him than from the auctioneer or owner, I approached him and asked about the New Holland Super 66 baler that was lined up in a field across the road from the farmhouse along with the other farm machinery.

"I used it last year. Made 4000 bales and never missed a tie. It's a good one."

When it came time to sell these machines, the auctioneers abandoned their public address system and left many of the people interested only in the household goods behind. A small group of farmers crossed the road with them to where the farm equipment was parked. Here the chant began as the machinery went up for sale. The baler was part way down the row of equipment, and when the auctioneer got to it, an initial bid of $1000 was suggested. No takers.

"Well, who give me 8,8,8,8?" No one would go for $800 either. Finally auctioneer got down to $200.

"200", I said. And it was mine.

The reason that we were able to assemble the necessary equipment to manage our farm so inexpensively was due to several events that changed farming in our region forever. By the spring of 1970 it simply cost too much for many of the smaller farms to meet the new requirements for Grade A milk. Cows now had to stand on concrete rather than on wooden floors, and refrigerated bulk tanks had to replace milk cans that formerly were set to cool in spring water. One by one as the hill farms stopped milking, the equipment from these farms was put up for sale. Meanwhile, the larger farmers who

153

upgraded their facilities during this period were purchasing big machines. The first really large diesel tractors arrived in our valley at this time. Although the large farms did not abandon their good small tractors, neither did they compete much for the smaller machinery coming from the farms that went out of the milk business.

Finally we were ready to make hay. I decided to cut just a few swaths in the River Meadow so that I could practice using the rake and baler before cutting a large area. The side-mounted mower snipped along, the serrated triangular sections of the knife slipping in and out of the finger-like guides that held the grass stems to be sheared. The reciprocating motion of the knife shook the whole tractor as the pitman arm driving the mechanism oscillated. It was going to take time before I learned just how far to move the lever that controlled the hydraulics in order to bring the mower blade up out of the grass at the end of the swath. I could see that some talent was required to keep the grass moving swiftly through and over the mower. If I let a wad of grass build up on the mower, it covered the mower sections and that part of the knife no longer cut. Lifting the blade and shaking with the hydraulics could sometimes dislodge the plug, but often a plug meant shutting down the tractor engine and climbing off to clean the blade.

In a couple of days when the hay was dry, I attached the hay rake to the VAC and went to the end of the swath. I lowered the rake so that the tips of the rake fingers just cleared the ground and moved the power take off lever to the "on" position. Instead of the smooth swish of rake teeth through the air that I expected, there was an awful clatter. Some of the teeth were hitting the stripper loops as the four rake bars cycled past. I turned off the power take off, and found the guilty strippers by looking for where they had been hit. By sitting on the ground and pulling hard, I could bend the strippers slightly so that the teeth would no longer hit them. I got back on the tractor and started it again. As I turned on the power take off there was more clatter. I had bent some of the stripper loops too far. I repeated the process several more times, but never got the smooth swish that was inherent in the implement design.

About that time, I noticed that the four rake bars had been broken and then welded back together. Slowly, it dawned on me. There were slight differences in the position of the rake teeth as a result of the repairs on the rake bars. It was impossible to adjust this rake to create the necessary space for clearance of every rake tooth. Depressed at this discovery, I got the rake set up as best I could and crept slowly down the swath. I had seen my neighboring farmers sailing down the swaths that they had mowed, their rakes revolving in a blur - dry hay spurting forth from the end of the rake, curling over itself like a perfect surfing wave. I crept along in first gear, the throttle just above idle, the clatter behind me threatening to become pieces of shrapnel each time a tooth hit an irregularity in the ground. I left a pathetic windrow down the length of the River Meadow.

I made a second and a third trip, my spirits sinking as fast as the afternoon sun. As I reached the north end of the River Meadow for the third time, I noted two figures approaching from where they had parked a blue GMC truck outfitted with a farm dump body. I had noticed the truck on the road before but didn't know who owned it.

"Saw you were having trouble. Thought we would stop to help," the older of the two said. The person speaking was Bernard Tullar who was with his son, Wayne. Thus began a lasting friendship with this family.

Bernard and Wayne pointed out that bending the strippers was unlikely to overcome the problem and suggested that I bend the teeth so that they would not interfere with the strippers. They described how this could be accomplished with levers fashioned from surplus iron pipe.

After finding and drilling some pipe, I returned to the rake in the late afternoon and bent the teeth. It was better, but on rough

155

ground occasionally a tooth would come down on the wrong side of a stripper. That was bad and what had probably broken the rake bars in the first place. My rake that seemed like good buy at the auction had turned out to be junk.

The next day, I tried baling. I set off down the windrow. The pickup fingers on the baler lifted the raked hay gently off the ground and into the baler. Under its guard, the connecting rod stroked to and fro as the piston compressed the hay in the chamber. There was also the clatter of other activity I didn't recognize, but I drove on. In a few moments there was a BANG! It was a knotter, pieces of which flew over my head and landed in front of the tractor.

In dismay, I picked up pieces of broken cast iron-lucky I hadn't been hit by any of them. The other knotter was encased in a tangle of hay rope. I had not realized that the mechanism that determined how often the bale would be tied by the knotters was set so that it tried to make bales only a few inches long. The extra noises I heard were the knotters working overtime. A needle had been deflected by the accumulating knots and had struck the knotter. The result was in pieces before me.

I put in a call to McLarens, which serviced New Holland equipment in addition to Case machines. I was amazed at how fast the serviceman arrived. Farm equipment dealers have learned that having hay down at the same time that equipment is broken constitutes an emergency and the dealers that stay in business react accordingly. Although this was a weekend, the service man arrived that afternoon and within minutes had bolted on a replacement knotter after telling me why the knotter had been destroyed.

A few days after my baler mishap, Bernard Tullar stopped by. "Stub Jenks has a good side delivery rake parked in the barn at the Sansbury place. He's never going to use it again. Old, but works just fine."

Even before I could call him, Stub's slow moving International Scout pulled up. Together we rode up River Road, past Stub's house.

The large barn behind it had collapsed a few years before, but the hay rake was in a neighbor's barn and had escaped destruction. We wheeled the hay rake out into the open. The A-frame drawbar was bent on one side. "Been bent for years. The first time we used it my son made too sharp a turn with the tractor."

It was easy to hook the rake to the hitch on the 9N, and, after greasing a large number of fittings, I set off to rake over the windrows that had been lying on the ground for several days. The rake floated along, hay cascading from the end of the rotating mechanism. There was no clashing or grinding, only a gentle "ting" when a rake tooth hit uneven ground and sprung back. I felt wonderful. One must go through an experience like this step-by-step in order to experience the joy that simple things can give.

Our mechanical problems did not end, but as we were learning a bit about animal genetics we were also learning how to deal with machinery - when to "ship" vs. when to repair. We acquired a number of additional pieces of equipment and all of these machines had an effect on us as we became accustomed to operating them. Some never felt right and with increasing experience we got rid of them. Others simply became an extension of ourselves.

As our cow herd expanded we needed more hay than we could raise on our place. We started to cut otherwise unused fields owned by our neighbors. Son Ross got permission to harvest a large field across the river. I returned from that field with the VAC going fast enough to make the front wheels shimmy. I wanted to make way for the traffic as I crossed the bridge. As I turned left onto River Road there was a sudden change in the note from the engine – a severe miss and loss of power. I backed down the hill onto the East Thetford Road and took the tractor with its sputtering engine to George Lawson's garage.

"Broken valve spring," said George over the phone when I called him later.

"Gosh, where are we going to get a new one? We have hay down and have to bale."

"Already fixed. Looked like a Chevy spring so I took one off the junker out back. Works fine!"

Happiness with a machine depends upon whether it is right for the time and the place. Each time I pulled the VAC hood latch and raised its bent hood to carry out routine maintenance I could look at that little engine that ran so hard for me and the farm, and I felt good about it. I would like to find the genius who specified valve springs on the Case identical to those on the Chevy 6 cylinder engine – probably the most common engine in America - so I could thank him personally. Acts like this have had more to do with America's success than most realize

All three children learned to drive the tractors before they learned to drive anything else. I could prop them on a fender with their feet braced against the gearbox or differential so they could observe my actions. After a bit of instruction, I put them in the seat and I sat on the fender to observe how well they had listened. The safety experts and risk-management lawyers in the farm equipment industry have a fit when they see this type of activity. In Europe, however, many of the tractors have a small seat built into the top of the fender for this purpose. In Eastern Europe there may be a couple of seats, one on each fender. A tractor there is not only used on the farm, but also to take the family into the town on market days. There are no warning decals on these seats. People learned to look out for themselves in countries where for centuries there was no legal recourse if the wheels of the king's coach left someone crushed on the road.

After operator training was completed, we generally frowned on having riders on the tractor. However in the rush to get some jobs done, a person would sometimes hop onto the tractor fender to get a quick ride to an operation in a remote field. I observed Ross driving a tractor at a good clip one day, a sister on each fender. He turned off River Road before the end of the steep bank and dropped sharply to the River Meadow below. He looked in turn at each sister to see their surprised expression and incidentally to see if they were still perched on a fender. He then caught sight of me and my frown. He was instantly sobered. Children caught at play - play with possible dreadful consequences. I didn't have to speak to him – he knew what was on my mind and turned his guilty face to the task ahead.

We later bought another tractor, an Allis Chalmers D17, big enough to pull the baler up any hill on the place – a wonderful machine. One winter day, I walked into the hay barn to do my morning chores. As I entered, I passed the gray rear fender of the 9N. When I bought that little tractor, I thought that it was all the tractor we would ever need or have. The knowledge that elsewhere in the barn there were now others, larger and stronger, was reassuring. What we were up to was a success. Our children were larger and stronger. Our relationships, like those with the tractors, were mostly happy - based upon trust and appreciation of what each could do for the others.

Chapter 11

Plowing

The Scene: Omaha, the McIntyre household in 1940.

While my brothers and I watched, my mother cut into the corrugated box with the bread knife and removed three large mugs from the package that had been shipped from Strawbridge and Clothier. It wasn't Christmas or a birthday. My mother's former college classmate had sent them to us for no obvious reason. It was just another example of the generosity and love that this well-off woman from the east offered to her dear friend struggling on the "frontier" of the west. The mugs went onto the window-sill above our kitchen table. One lunch after another while we ate our peanut butter sandwiches we read through the glaze to the poem painted on the mugs:

> Let the wealthy and the great
> Roll in splendor and in state
> I envy them not, I declare it
> I eat my own lamb
> My own chickens and ham
> I shear my own fleece and I wear it
> I have lawns, I have bowers
> The lark is my morning almanac

160

So jolly boys now
Here's God speed the plough
Long life and success to the farmer.

Years later, long after those mugs had been lost or broken, a wonderful office manager who worked with me at the Cancer Center, presented me with a package. I opened the box, and extracted a mug from the packing materials: "Let the wealthy and the great…" It was identical to the mugs that had been on the window casing in Omaha.

The office manager was a Philadelphia girl, and one whose mother purchased things at Strawbridge and Clothier only for very special occasions. She had been given this mug as a child, and she knew that even as I struggled with our emerging Cancer Center I was also trying to keep a farm going. So I was touched by the gift - still am. It resides in safety on my bureau today.

God Speed the Plow. Listen to the music and the voices on the net. A great song if you have a cup.

I believe the song was written by an English romantic who composed the original verses. Since then, many additional verses have been added, often by those enjoying hearty fellowship and enough drink to call forth a song. I am sure an occasional fellow would have chimed in as he walked behind a plow. He would have worn a leather apron, and flipped the plow over at the end of the furrow. Each stanza lasted for about one furrow length. That is why there are so many stanzas and why the plow would not cut the soil in the turn. The men over at the East Thetford Commission Sales would have thought the "jolly boys" wording used in the invocation silly. Those with their hands near the earth, however, would have no trouble appreciating what the poet was trying to say. Maybe reading those words so many times led to my family eating its own lamb and fetching our own eggs.

*****

I was plowing the River Meadow. As our County Agent, Dick Rutherford, had said, fertilizing our pastures would give us more of what we already had in the way of weeds and grass. If we wished to change the composition so as to include alfalfa or clover it would be necessary to plow and seed.

My tractor and the furrow that followed it wobbled down the field. Although the loose steering on the VAC was partly to blame, the real problem was a driver who simply lost sight of his target, corrected and then over corrected. After the first furrow, there is nothing that can be done to straighten out the job. On the next trip the wheels of the tractor are locked in the groove created by the cutting of the first furrow and every wobble in that line is transferred to the second line, etc. While I was making this mess people driving by on River Road could look down into the River Meadow. Many of them could have done a better job. Instead of ridicule, I received friendly waves. I returned the waves, sometimes imparting a new wiggle to the furrow while doing so.

Soon I resigned myself into having a field that was full of twisted furrows. I relaxed a bit as I became used to the pull of the plow on the tractor. I could sense the snap of innumerable roots as the plow sliced along - a slight vibration transmitted through the steel of the plow and the castings of the tractor to my feet as my friend the VAC and I slogged along. Without the staccato of the tractor exhaust I might even have been able to hear the roots being stretched and cut – millions of rubbery guitar strings. I thought about the small amount of gas that the tractor used and how much earth the plow had already turned over. I put myself back in my parent's garden, a garden fork in hand, turning the land over forkful by forkful. My back ached at the thought and I assigned a higher priority to that precious essence of petroleum slowly draining into the carburetor – gasoline consumed molecule by molecule, transforming sod into seedbed.

Before long, my furrows started to creep up the slope toward River Road. As the hill grew steeper, the front wheels of the tractor started to slip sideways, the tug of gravity being too much for them to cope with. To stay on track I stood on the brake for the left rear wheel and turned the steering wheel a bit to the more to the left. Then I noticed that the strip of earth being lifted by the plow was no longer being turned over. Instead, the strip paused, rested a while on edge, and then fell backwards to its original position.

I stopped, got off the tractor, and tried to arrest the falling line of sod by hand. The huge weight of just one furrow was soon apparent, as was my inability to cope with this new development. I adjusted the plow and tried again but things were no better.

I imagined that the harrow could smooth up my mess, but Bernard Tullar told me later that all the harrowing in the world could not make up for a bad job of plowing. He was not being critical – just pointing out a truth.

Henry Ford knew his small tractor would be a non-tiring surrogate for two hungry draft animals. On such a tractor, the driver sits close to and can smell the newly turned earth, can easily see the exposed earthworms as they wiggle to find cover, and can hear the flap of wings as the birds land for the worm and grub harvest only a few feet behind the tractor. The machinery of today is different. From high in a sound-insulated cab the driver never senses the snapping of roots. The romance is gone, but food is cheap because one worker does the work of many, and chemical herbicides now allow many fields to be planted without first plowing them.

Eventually, I reached the point when the slope was so steep that it was folly to continue. I did not wish to flip the tractor. The fellow who last plowed the field had been walking behind the plow, not riding a tractor, and I had to give up. The furrow where I stopped remains to this day as a depression on the surface of the River Meadow.

I moved down to the next section to be plowed, created a dead furrow and lined up my furrows on both sides. At the end of the afternoon, I had completed about six acres, a ragged job - something like the first haircut that Jean had given me. Now we needed to break up the turned over sod. For this we had two disc harrows, one to be pulled by the VAC and a smaller one, pulled by the 9N.

A tractor pulling a disk harrow is very much like a ship dragging its anchor. In order to make progress the prime mover must continually overcome the unrelenting resistance to movement by the implement. When the harrow is correctly set up, it is all a tractor can do to move the harrow fast enough to churn the soil. Jeanie and Elizabeth took over many household chores so that Jean and Ross could harrow each afternoon. As I pulled into the driveway after work a couple of days later I spotted Ross on the 9N far down the field, the little harrow bouncing along behind. Despite the knack the 9N had for burying its mower drive in the furrows and bogging down, Ross almost never got stuck. Not bad for a 10 year old.

Near the end of the week, the Agway truck arrived with a load of lime. With extra large "flotation" tires to keep it from sinking in the soft dirt, the truck careened around the field. As it bounced along, a cloud of white particles enveloped the rear of the truck. I watched from a distance, my skin stung from an occasional hit by an outsized particle. A short time later, the truck returned and spread 500 pounds of fertilizer per acre.

On Saturday, George Evans came over from East Thetford. A seeder was behind his tractor – an old timer with high wooden-spoked wheels and iron oxide paint. When we opened the hoppers to pour in the seeds, I was impressed by how new the insides looked. George had cared for this piece of equipment well. Earlier we had tossed the alfalfa seed with a bit of corn syrup to make it sticky, and added a powdery bacterial inoculum that adhered to the sticky seeds. These bacteria would form root nodules on the alfalfa, nodules that would take nitrogen from the air and fix it into a form that would be used by the growing legume. Now we tipped the alfalfa seed into the small hopper, and a bag of oats, the cover crop, into the larger hopper. As George drove off in a cloud of dust I had confidence that the care he had taken in maintaining this old seeder would be duplicated in performing the seeding. In a short time he was done. When he came up to the house, he looked at the northeast window of the first floor, the window at the end of the living room where the Squire's wife, Clef, had put her piano.

"As a young man I stayed a winter in that room," he said. "I spent most of my time cutting wood trying to stay warm. If you look on the floor next to where the stove was you will find my tracks." A couple of years later we scraped away the thick layer of battleship gray floor enamel that the Squire had put down. Right next to the chimney in that corner of the house were the deep wounds made by George's axe. Coals from the stove had charred well into the floorboards in a couple of spots. Houses that survive 200 years have many narrow escapes.

If the five of us had walked the field in a line broadcasting the seed (as we did when reseeding another pasture), we would not have needed George and his seeder. We would have used much more seed, however, and would not have had a nice even seeding of the crop. We also would have missed out on the story concerning the axe marks. From George's tale of chopping wood I realized what life in the 1920's must have been like on the farm. Much of the space in the house, space that we took for granted, had housed hired hands necessary to get the major jobs done. The kitchen that we found so spacious had been crowded with hungry men around the table, and during winter they each tended a stove in their assigned space. That downstairs room George used didn't need a closet. His few clothes hung from the square nails in the wall.

Until stoves replaced the fireplaces in each room, even more manpower would have been required to cut wood - to survive the winter without freezing. Our lovely brick house, the beneficiary of Squire's remodeling, was an early example of "gentrification". The darker history of a brutal farm life in a cold house might have been lost but for George's visit.

George's seeder had drilled the alfalfa seeds into the prepared earth and flipped a bit of earth over them. He returned later with a home-built roller behind his tractor and compacted the soil a bit to hold the seeds in place. Soon, the oats sprouted, and protected by these vigorous stems, the delicate alfalfa plants emerged. A few inches beneath the surface the overturned sod fermented giving up its nutrients to the new plants as it rotted away. The acid soil, the curse of our region, had been made sweet by the lime. The amounts of nitrogen, potash and phosphate were balanced optimally for the new seedlings, and rain, the one advantage conferred on New England farms, was gentle and well spaced.

The results were phenomenal. The oats shot up and soon headed out with immature grain. Nestled amongst these oat stems alfalfa, young and delicate, flourished. Soon we would mow the immature

166

oats taking a gentle first cutting of the alfalfa at the same time. We would make our first really good hay.

In the next couple of years our plow, now pulled by the larger Allis D17, turned soil in the other fields and a similar transformation occurred. Fields of thin stemmed grasses decorated with Indian Paintbrush gave way to tangles of alfalfa, aromatic purple clover, white clover buzzing with bees, and the waving grass heads of timothy. We stood in freshly cut hay, clover almost up to our hips while we unplugged the teeth on the mower, the air permeated by the sweet smell of blossoms. When we got behind in our work and the timothy heads matured before cutting, we saw the crop reseeding itself. There were times when the forage grew too well. We bought a better mower and mounted it behind the VAC. Even with the new mower, there were days when we spent as much time taking tangles off the mower as we did on the tractor seat.

Our harvest depended upon the top few inches of soil that we turned over with the plow, fertilized and limed. Although some roots pushed downwards into the subsoil in search of moisture, the conversion from the sparse collection of native grasses and paintbrush to the carpet of dark green, was dependent upon nourishment available in those few inches of topsoil. This was made clear to me when I plowed the Red Cow Meadow.

Several months earlier as the cows ate the thin grass and the brush along the edges of the adjoining Hemlock Pasture, their appetite producing a golf course-smooth surface, I noted a shallow ditch following the contour along the hillside. I followed the ditch as it led up hill and into the White Gate Meadow. From there it entered the woods along Ross's brook. I could see that it had once intended to pick up water from the brook and carry it to a point where the ditch went under the stone-wall at the top of the Red Cow. Some large pines, maybe 100 years or so old, had grown up through the ditch at its top end. These must have grown up in the pasture long after it had been cleared. This indicated that the ditch was an old feature on this landscape, at least 150 years old, and likely going back to the origin of the farm. It is unlikely that it would have been used for irrigating the Red Cow. The forty inches of rain we get in an average year makes that unnecessary. It must have taken two or three weeks for a man with shovel, wheelbarrow and pick to dig that ditch. Lots of effort. I walked along the ditch several times puzzling over its use and finally gave up.

Now, as the two-bottom plow sliced through the sod about half way up the gentle slope of the Red Cow I suddenly heard some loud clanking. The plow was turning over soil laced with brick! I stopped and got off the tractor. These bricks were all imperfect. Many were "burned". They had gotten so hot in the kiln that they had melted, turning black and deformed. Others were broken in two, were warped or otherwise imperfect. I stared at the area in disbelief. I was in the middle of a brickyard – a place where bricks had been molded from clay and baked in a kiln. Some of the soil around this area was heavy blue clay, perfect for making brick. I had never questioned why that slope of the meadow contained a gentle alcove, but now it made sense. That is where the clay had been mined for the brick. And the ditch? Water was needed to moisten the clay and it is a lot easier to move water in a ditch than in a bucket. The bricks that Ebenezer Green used for his house were made on this spot, not hauled by horse and wagon from some distant brickyard. The brick in Clyde Grant's house half a mile away, and those in a house two miles away on the North Thetford Road might also have come from this brickyard. Everyone, even people in timber frame houses, would have needed brick for chimneys. Ebenezer Green had run a brickyard. I suddenly felt very close to the man.

After the brickyard had been abandoned, grass grew up in the cracks between the rubble left behind. Over the years, the decaying grass on top of the brick and the accumulated casts from innumerable earthworms buried the brick debris. Bit by bit the evidence of the brickyard had faded. The dead furrow in the Red Cow was recent enough to have been made by a Steele, however. I wondered whether he was surprised as he walked behind his plow to find himself in the middle of the former brickyard. Or had he learned from tales coming down through his family about Ebenezer Green's enterprise? At any rate, those bricks had been buried at least once for Steele and once for me to discover.

When fields are limed, fertilized and seeded with hybrid corn the resulting crop utilizes an amazing fraction of the incoming sunlight to generate food. We take this productivity for granted. It is why we can afford to abandon less fertile fields. We don't need them anymore. Fields that once held corn are given over to pasture. Less good pastureland reverts to forest. Abandonment is seldom planned, however. It just happens over time.

Bernard Tullar told me of the care with which farmers in his father's generation mowed. Scythes swinging easily in their hands, they mowed right up to the edges of the stone walls and took the hay. The contestants today in scythe mowing events move through high late-summer grass that has been saved out for the competition, bright sun streaming in, dew long since gone, and are judged on how and what they have cut. The onlookers, once they have watched the rhythm of the mowers in the summer heat, should then go to Robert Frost for the rest of their instruction in mowing:

> "I went to turn the grass once after one
> Who mowed it in the dew before the sun.
> The dew was gone that made his blade so keen
> Before I came to view the leveled scene."[1]

Robert Frost's mowers could leave a tuft of flowers and the meadow lark's nest intact, while cutting to the stone wall. Today, it is still possible to swing a scythe while greeting the first rays of sun.

---

1 The Tuft of Flowers. Complete Poems of Robert Frost, Henry Holt and Co. 1957

I gave it a try to see what it was like. The sunlight burns through the fog on a 40 degree July morning to illuminate something magical. Grass and flowers encrusted with dew liberate a shower of tiny diamonds that hover in the sunlight for an instant before settling on the fallen hay. The beaded spider webs beg for closer inspection, and the crisp air is so full of fresh perfumes that the lung rebounds from the insults of the last winter. As the sun grows stronger, the wet denim against the mower's leg cools the heat of exertion. The scythe is raised, the snath placed against the ground and the blade brought to the eye. The hand strips the wet grass from the blade and the sharpening stone is fetched from the pocket. The trickle of cold water that flows from the blade onto the hand feels almost as sharp as the blade itself. The morning birds sing.

The romance of such mowing, however, is tempered by the presence of poison ivy along the edges of the field and rocks that dull the blade. Worst of all are tendrils of old barb- wire that have lain hidden in the brush until they seize the blade, jerking the shoulder of the mower and ruining the edge. Better to leave these hazards for the teeth of hungry sheep to sort out.

There are no longer enough sheep to patrol all the stone walls. One by one the fields close in, and the stone walls now outline sections of forest, the maturity of trees in these sections governed by the time that the field was "let go." Meanwhile, down in the valleys the plows (or, now large "transport" harrows) slice along, turning the soil so that in the fall the harvest of the summer sun can be dumped, still smelling of corn pollen, into the silage bunkers.

Chapter 12

Haying

Haying was usually a series of three-day events. We mowed on day one. The mower attached at the rear of the Case was followed by the Allis D17 that powered and pulled the hay conditioner. The tractor exhaust spouted skyward as the Allis power takeoff spun the conditioner roller that crushed the grass stems. Each crimp in the stems could now bleed moisture to the air. On day two, we waited for the dew to dry and then tedded. The whirling teeth on the tedder tore apart the clumps of wet grass that had tangled in the mower blade. The stems that had not seen the sun the day before were now on the top of the swath where they would dry. On the evening of day two, the dew would again make the hay limp with moisture.

Three sunny days in a row. That is what we hoped for, and in a New Hampshire June this is hard to find. Hard enough so that not too many years ago, farmers did not try to make hay in early June. This changed when it was shown that early cut hay, before the grass went to seed, had more nutrition. Well-cured early hay should have twenty percent or more of available protein in it. Cut later, it might have less than half this amount.

By the middle of day three we were in the field, picking up the hay crushing it in our hands, judging if it was dry enough to be baled. The rake, greased during the early morning fog when little else could be done, would be poised at the edge of the field, waiting to go. At last, when the hay was dry, the 9N pulled the rake along in third gear throwing up a cloud of dust as the tines tumbled the hay into windrows. The rake brought any remaining moist hay up off the ground and into the drying breeze. There the flattened and crimped stems would bake to crispness in the sun. An hour later, the Allis pulled the newly greased baler onto the scene, and soon the pounding of the wadboard and ram echoed from the hills.

There was a bit of exhilaration as one stood on the deck of the Allis to get the best view as it pulled the baler over the fields. The River Meadow gently unrolled ahead of the tractor. To the side, one could watch the fingers on the baler pick up and carry the raked hay into the auger. I sometimes felt that the machine was part of a larger stronger me - those were my fingers picking up the hay. I felt the increased effort of the engine when the windrow became thick and tangled, and I pulled the hand clutch to cut the tractor speed by1/3 while also increasing its power by the same amount. The ingestion of the thick hay slowed and I listened to the engine catch up with the load. It was like listening to my own heartbeat. Maneuvering around obstructions with large equipment I sensed that my own body extended to the periphery of the machine, Body English was coupled intuitively to the steering wheel and a sense of accomplishment flowed from a clean line of travel we left in our wake. I suspect young Ross had similar thoughts as he took charge of the baler. When it went well you really felt good about what was happening.

Ross at the bottom of the hill in the Red Cow Meadow on the Allis D17. It pulls the baler that is behind and out of sight as he cleans up hay missed on the first pass.

All of this activity depended upon the continued presence of the sun. Once the sun dropped, the dew was not far away, and as soon as the dew was on the windrows the baling was finished. In the bright sunlight – the heat of mid-day - the hay was crisp and judged ready to bale. By the time the bales arrived at the barn, they never seemed as dry - whatever moisture remained in the hay was now more apparent. In fact, hay that was absolutely dry in the field was easily shattered by the tedder and rake. We didn't wish to have it so dry when we worked it that much of the nutrition would be lost in the leaf powder that blew away.

I remember one day - a day that was supposed to be hot and sunny - when, as I pulled the tedder along behind the Allis, I could see the vapor trails from high flying jets in the sky to the south.

Instead of evaporating into the blue as usual, the trails persisted, joined later by those of other jets flying the same route. Coalescing, these formed a thin blanket that slowly moved northward, finally coming between the sun and me. My skin immediately sensed the loss of the driving heat that had been there moments before, and as the jets continued to fly and the sun remained veiled, I reluctantly put the tedder away. No point in making hay on that day. It wouldn't dry.

Although I was angry at the airlines that owned those planes and the people who rode on them, I realized that this was immature. Those people, five miles up, could not see me - did not know that I was down on the ground driving the tractor to the barn in disgust. My problem was not the airlines or passengers, but my attempt to make hay during a long-scheduled vacation from the medical center. I had to make plans weeks ahead of time, guessing at the weather and recognizing that June was often a wet month in New Hampshire. My anger derived from trying to do a good job at two tasks that were often mutually exclusive.

As time went on, I spent less time shaking my fist at the sky and more time making certain the equipment was ready to go when the weather turned right. We noted that Walter Record often was able to get his haying done when we had procrastinated, awaiting better weather. Soon, we cut when he cut, and if it rained on the downed hay, at least Walter's was getting wet too. Eventually the children were large enough and strong enough to take on much of the haymaking and my schedule became less critical.

When the afternoon sun dropped behind trees at the edge of the field, the hay in the shadow became damp with the first dew. The bales from this portion of the field were heavy. "Careful, this one has an anvil in it!" This heavy bale would go on the floor of the hay wagon – too heavy to lift far. Tightly packed wet bales provide ideal conditions for the growth of molds. Spores for these molds float through the air, land on wet hay in the bale and begin to grow.

As the mold grows it produces heat. It is not long before such bales become warm to the touch, even when they are left out in the field to be cooled by the breeze. Heat can build in stacked hay for quite a time, sometimes days, until temperatures deep inside the hay are well above those required for ignition. When the heat finally arrives close to the surface where enough oxygen is present, the ignition may be nearly explosive. Barn fires often erupt with little warning.

When confined in a stack, the lack of cooling air allows a moldy bale to become hotter, and, up to a point, the hotter the bale, the faster the mold grows. By the time the hay in the bale is finally too dry for the mold to grow, much of the nutrition in the bale has gone into sustaining the growth of the mold. As the mold dies in this now-dry bale, the spores form. When the bale is finally opened, these are released in a cloud of smoky dust comprised of billions upon billions of particles that float away, some of them to eventually land upon the next wet bale.

Recent advances in farm machinery have revolutionized the harvesting and storage of hay. Some balers produce large round bales weighing six-hundred pounds or more. In wet climates the hay in these bales may be wrapped in waterproof plastic for outside storage, an alternative to costly hay storage barns. Other round bales may contain damp hay that becomes silage as it ferments in the wrapped plastic. A lack of oxygen in the wrapped bale allows the bacteria in the damp hay to grow and make silage (organic acids that preserve the nutrition and prevent mold growth.) Cows love it! The specialized equipment required to produce these large round bales is expensive, putting this method of haying beyond the reach of smaller farms. So thousands of square bales still go into the old hay barns that dot the landscape. Among these thousands of bales, especially those prepared with farm machines that reduce handling by humans, there may be a few that are far too wet to behave in a benign manner.

The rush to get bales under cover, and the presence of many helpers on a busy day may lead to carelessness in excluding overly damp bales. Our hay elevator allowed us to move some of these heavy bales into places where they should not have been. When they came off the elevator we attempted to put them on the outside of the pile we were building where they would be cooler. If we had depended upon our muscles for lifting them up the pile they might well have been left on the ground where they belonged.

The Tullars purchased a much longer hay elevator. They installed the many sections of their elevator so that it lifted the bales through an opening high in the gable end of their barn and traversed the huge hay mow inside. At various points along the top of the mow, the elevator was set to tip the bales into piles below. This saved the labor that was required otherwise to lift and then distribute the hay bales within the mow.

After several years of experience making hay on our farm, Ross started to work on the Tullars' farm when he wasn't busy at our place. Because the elevator easily handled wagonload after wagonload of hay, Ross was pleased with this arrangement despite the occasional need to climb into the mow to adjust the mechanism.

One hot summer day Mike Smith suddenly stopped his car on River Road and shouted to us, "Parson's barn is on fire!" I looked to where his finger pointed and saw a black smoke cloud rising above the trees to the north of us. In a flash, I joined the rush of traffic moving north in the direction of the smoke.

Overarching branches along River Road interfered with the view ahead and it took me some time to realize that the smoke originated north of the Parson place. The likely source was the Tullars. When the traffic became too crowded to continue further, I pulled off the road and sprinted toward the Tullar barn. From the shell of the flaring barn a tremendous black cloud rumbled skyward. My mind was on Ross and the possibility that he was in the midst of that inferno.

176

From somewhere came a voice, "Ross is OK." To this day I don't know whom to thank for that information. A volunteer fireman directed me to the river where a suction hose connected to a pump started the water on its way to the fire. I was given a smaller hose and told to use its jet to divert any debris that could get into the suction hose and clog the pump. From my position down low next to the water I couldn't see what was going on at the fire. I could hear more fire engines arrive and the heavy crash of falling timbers. Minor explosions tapered off, and the black cloud faded in intensity. Large fragments of debris continued to be swept into the sky, however. Eventually, I was relieved from my post and crawled back up the bank to the road.

Ross and I found each other. The farm crew had been loading hay into the barn when the fire broke out. The cows, fortunately, were out of the barn, and no one had been hurt. Ross had run to remove the Tullars' large Allis Chalmers tractor from the barn but could not start it before he had to escape the heat. Minutes later the barn and its contents was gone. Our brief conversation was held against the backdrop of a steady spray of water that was preventing the heat from igniting the farmhouse.

As the day finished, the cows were being hauled to Bernard's brother's farm to be milked. A black mass, still hissing as the fire hoses hit it, was all that remained of the summer's haying. It was time to start over at the beginning. "Jolly boys now," indeed!

Only a few years later, Shine King's barn was consumed in flames. From my secure vantage (by now I was a faculty member with tenure) I didn't understand how people whose livelihood went up in smoke could take the blow. They have to be strong.

*****

During the haying days of summer, a gentle breeze of cool dry air from the northwest is pleasant. Wind from this direction, however, often means that rain is coming soon. What we really wished for when we hayed was a strong, hot southwest wind, a wind which came across miles of the dry heartland of the U.S. gusting through our windrows and drying them fast. Sometimes during the second or third cutting and with such a wind blowing we could make hay in two days.

Instead, our air was often stagnant and humid, a result of the "Bermuda High" - air under a sun partially obscured by haze, ending in an evening that brought "heat lightning" and sometimes violent thunder storms. In this weather the grass and clover retained the damp at the base of the stems throughout the day. Whereas dew assists the mower wielding a scythe, damp stems clinging together soon tangle a mowing machine.

On one such day I was mowing the River Meadow. Each time I came around the north corner and headed south I could see the field stretching before me, and I figured that this time I would have a clean shot to the far end. Instead, before I had gone far the mower was tangled. I hopped on and off the tractor five or ten times in covering that ground - my sweat dripping, my straw hat limp in the humid air. Far ahead, our neighbor George Hobbs was up on the roof of a new porch he was building. As he pounded shingle nails under that hazy sun, I could see the air above the roof ripple in the heat. "The only fellow in Lyme who is hotter than I am today," I thought. I struggled past his house and continued my frustrating trips around the field hoping that eventually the mowing would improve, although there was little cause for optimism. A trip or two later, as the VAC approached the fence next to George's house, I found him standing in the shade of a small tree and leaning on a fence post. On top of the post was a cold can of beer, and George had another one in his hand. I stopped, and we stood, one on either side of the fence, commiserating with each other about our jobs and enjoying a beverage given to mankind by God for hot days.

Several years later, George developed acute leukemia and I became his physician in addition to being his neighbor. As the bonds between patient and physician developed and as George fought this most difficult of diseases, the gift of that cold can on a hot day underpinned our relationship. He had a good remission, but eventually his disease returned. His funeral was held on a cold gray day of winter. Returning home afterward, I stood in the study of our brick house and looked down River Road to the roof that George had shingled that day. I thought of that meeting at the fence as I listened to the crackle of firewood in the stove, and I stared long into the glass of Scotch I held.

Making hay the way we did offers plenty of opportunity for employment of strong children and volunteers. It also goes better when there is plenty of hay-making equipment. We were lucky that Mike and Jean Smith wished to cut hay for their horses and we could share equipment. Their tedder was the best design available and their hay wagon was larger than ours. We were all available at various times for picking up bales. Sharing the job helped get the hay in the barn before the rain came.

One young physician in training at my medical center treated his guests, visitors from the city, to the joys of haying. Loading them in his car on an afternoon when it was likely haying was underway in the county side, he would drive about until he saw haying in progress, stop the car and he and his crew would run into the field to help. The farmers, often dumbfounded by the appearance of volunteers, never turned them down. When bales are on the ground and an empty hay wagon is parked nearby little instruction is required by the volunteers.

We were always grateful for such help. One day I looked back from the seat of my tractor as I pulled a hay wagon and found a retired physician putting bales on the wagon. He and his wife had appeared out of nowhere. Soon he was on the wagon stacking bales that his wife and others threw aboard. Here was a fellow whose career had placed him at the forefront of revising the medical curriculum used in U.S. medical schools who found interest and fun in tossing bales of hay. We finished picking up the hay in the Home Meadow and paused for some lemonade at the picnic table behind the house before going to bale the small amount of hay remaining in the River Meadow. When we moved to the front of the house on our way to the River Meadow, we found that the road was wet. During our lemonade break, a sneaky and very local rain had fallen on our hay without touching our lemonade party. Our unexpected volunteer had encountered unexpected events in his drive to revise the medical curriculums. Haying was no different.

Those who often make hay learn to interpret the various clues that indicate what the local short-term weather is going to be, and can adjust their schedules accordingly. One day when we could see a storm coming from a long way off, we filled the hay wagon fast and had to empty it before we could pick up the remaining bales. As we drove toward the barn, it became clear that we would never have time to unload and get back to the field before the rain arrived. Just then, Bernard Tullar appeared with his tractor and an empty hay wagon. "We just finished, and I figured that you might need an extra wagon." We pulled our wagon into the hay barn, loaded Bernard's and got it into the large bay in the cow barn just as the rain began.

The hay piled in the two large bays of the hay barn was about three stories high by the time we had stacked the 40 to 50 tons we required. When we unloaded hay wagons, especially when the stacks rose above our heads, our sweat-soaked bodies were soon covered with itching chaff. The dust coming from even the cleanest bales roughened our throats and noses. When the last bale was stacked, we would head for nearby Post Pond. By the time we got there the sun was long gone, the moon was up and the water was often warmer than the night air. As we swam, the sweat and the itchy chaff were left in our wake. We dunked our heads and the hay stems left our now pliant hair. Soon our noses and throats were restored as well. In fresh clothes or swathed in slightly damp towels we piled once again into the pickup and headed for our brick house. There, the box fan in the attic window drew the coolness of night through our bedroom windows and the light of the summer moon illuminated the newly groomed fields outside.

Chapter 13

The Departure

The years 1976 through 1978 occupy only one page of our farm journal, a record that I started in the first days of our time on the place. This is not because so little happened during those years, but rather that so many things were happening that we couldn't find time to record most of them. I had started the journal so that we might have a durable account of what we were up to. Now that our farming operation was at its peak, we were too busy to leave an adequate record of what was going on.

As my professional life became ever more complicated, I accepted it in the way I had accepted my internship schedule so many years earlier; it was part of getting from here to there. The tremendous amount of farm work that Jean and the children took on camouflaged the growing limitations on what I could contribute to the operation. But it was only when some unexpected, often bureaucratic, blunder called me in to the medical center during my time off that I felt a tightening straightjacket and became stressed.

Most of the time I coasted on, because I was having a wonderful time both at work and on the farm. Several events coincided to cause significant changes in the routine. Jeanie had reached sixteen, and knowing that it wouldn't be long before she and the other children would be fully independent, Jean and I had resolved

182

to make a western trip with the family. I managed to schedule two weeks of vacation from one year back to back with two weeks from the forthcoming year. Off we went. Not before selling some cattle, however.

Our herd had peaked at 27. Before planning the trip we had already sold 8 cows and calves. Now we sold 15 head, butchered one, and kept our boss cow, Annabel, and two others. This simplified the management of the place while we were gone and reduced the amount of hay we would have to harvest that year. Seven of the cows went to Tony Farrell, who owned a fine field on Route 10 on the way to Hanover, a part of a former dairy farm.

The trip west was wonderful. The children thrived - each took on appropriate tasks as we packed up and made camp each day. They were enthusiastic about the landscape that unfolded before us, and were especially interested to see how barns, machinery, and crops changed as we crossed the country. It was a very happy crew that returned 4 weeks later.

On our return we found a message from Tony Farrell. He needed help with the cows that we had sold him. The field he had put the cows in had a stout cow-proof fence along the highway. The fence across the back of the pasture was in poor shape. Apparently our cows, like most stock when placed in a new field, walked the perimeter to learn the lay of the land. They walked right out through the back fence into the leafy hardwoods. They meandered about, stripping the sweet leaves and leaving a maze of tracks that said that cows had been there but not where they were going. Behind Tony's field, country lanes cut the forest and abandoned farm land into rough chunks all the way to Moose Mountain about 4 miles away. The cows were not hesitant to cross roads.

As in a totalitarian state where the dictator is suddenly deposed and leaves a power vacuum, the cows without Annabel split into several marauding alliances, roaming the countryside in contentious groups of ever changing composition. Unfortunately, a number of impressive country houses were scattered throughout this backcountry, most of which had fine gardens. As Tony's new cows roamed during our month's long vacation, some people had taken to sleeping out with loaded guns next to what was left of their vegetables. But the marauding continued. Tony was trying to contact a man with cow catching dogs to run the herd to ground, but hadn't yet been able to set this up.

We offered to help in catching the cows. It was not long before we were summoned to a long-overgrown hilltop pasture alongside Goodfellow Road where the cows had been spotted. We arrived soon after the call and waded into the brush, several of us carrying ropes and buckets of grain. Walking slowly and quietly we saw the distinctive Hereford color in the midst of a dense stand of small trees. I called to them in a reassuring manner, rattling the grain bucket as I did so. In an instant they turned, blasting through the brush, leaving a trail of destruction in their wake. They were gone. I remember thinking, "My, they look healthy and well-fed."

We moved on, continuing with the grain bucket rattling and friendly calls. A rain began and soon it was pelting down on the leaves so hard we could not hear each other's shouts. Then, we spotted the cows again, moving fast and splitting up. I fixed on one and decided to stick with her, which I did for a few minutes before she galloped off into the trees and I lost her. Water poured from my jacket, and sloshed in the grain bucket. I dumped the grain on the ground, thinking that cows might come back to this spot if we left enough grain out for them. I looked around. Where were the cow chasers? How far had I pursued that last cow? Which way was the road? I shouted into the tumult caused by the downpour. No answer. Shouted again. No answer. Solid gray overhead. No sun. I looked at the lay of the land trying to find the direction to the road. I realized once again that my Nebraska upbringing left me unprepared for direction finding on a rainy day in dense forest. I looked for stone walls and fences. Couldn't see any. I looked for moss on tree trunks. No help. I looked to see what direction the hardwood trees leaned, hoping that I could find south by this technique. Better in books than in a mixed hard/softwood stand in a downpour.

I was stumped and took the only way out, a steady downhill slog through the woods, to a trickle that became a minute stream, then past a stone wall and a heap of moldering boards, a former sugar house to a larger brook. Another quarter of a mile and I heard a car on the highway in front of me. I could not for a moment believe that I had arrived at Route 10. I had figured that I would emerge two miles to the east. I was one mile and a hundred and eighty degrees off.

We gave up on the cows that day, but Tony returned to put out more grain each day. The cows returned to eat and finally after several more days he corralled them.

After the cow chase, things quieted down for a couple of weeks while I caught up on work not done during our vacation, and prepared for a trip to the International Hematology Congress in Kyoto. I worked long hours with no chance for exercise. During this time I developed leg pain of the sort that can come only from a ruptured vertebral disk.

On the way to Japan I sat for hours in a coach class seat and the pain became worse. Then it gripped me when I stood as well as when I sat, and then, eventually even when I lay down. I returned from Japan, suffered through an important committee meeting at the National Cancer Institute in Bethesda, Maryland, while the sensation of having a large Police Dog eating on my right buttock grew ever worse. Finally, I gave up and saw my friend, an orthopedist, expert on back pain.

After six weeks of bed rest I went to surgery. As I awoke from anesthesia my back felt like I had been hit by a truck, but the pain – the dog-biting pain – was gone!

I resolved to be more conservative about how I used my back in the future. However, merging life on the farm and at work became progressively more challenging. And there were times when things on the farm just had to get done. We had fewer cows now, but the sheep flock was growing fast.

Jean was convinced that loading a heifer into Mort Bailey's truck caused the reappearance of my leg pain three years later. Who knows? I had done plenty of other dumb things after my surgery without apparent harm. I felt fine and the surgery seemed ages ago. Anyway, my symptoms had returned, I was on bed rest again. A few weeks earlier she had had surgery on the knee that was injured when her horse threw her. After her physical therapy sessions, she usually stopped at the food store in Hanover, picked up a bag of frozen vegetables and put them on her knee as she drove home. We were eating a lot of frozen vegetables during those days. I was lying on the sofa in front of the kitchen stove one day when she came in with the mail.

There was an envelope bearing the return address of the real estate agent who had sold us the house, Bob MacDonnell. Inside was a 35mm Kodak slide – a photo of our house. Much better than the usual real estate photo it showed the pleasant façade against a fine sky, and conveyed the sense of a special place. There was a handwritten note along with the slide. "I have many buyers looking for a house like this. Do you know of any that might be for sale?"

Jean read the note and looked at me. Only moments before, relegated to confinement on the sofa, I had wondered how I was ever going to find the time to repair the glazing that held the glass in the thirty-seven windows of the house. Putty was dropping out, and it had to be done before they could be painted. I returned Jean's look and at that moment our decision was made.

In my professional life, I made a lot of important decisions. At these times I drew upon an analyses of reasons for and against making a particular choice, gathered information that was not always easily discovered, and proceeded in a deliberate manner. When it came to my personal life, many decisions were rendered on the spur of the moment, for instance, this decision to sell the house. Perhaps careful analysis had taken place in the subconscious realm but there was no organized index to it, and the detonator - the thing that made something happen was something unimportant - for instance the colored slide of the house that Bob had sent to us.

A couple of years earlier, I had been offered a fine job at a medical school in the south. The position was right, the people were right, but I procrastinated. I had returned, still thinking seriously about the position, to cold New Hampshire and slipped back into the routine for several weeks. In the morning I went out to toss hay to the cows. It was cold and the wind came from the north, whistling between the vertical pine boards on the hay barn. I looked at the whitewashed walls around the stanchions, heard the wind once more, looked at the cow tongues sweeping our good hay into grateful mouths, and made my decision. I went into the house and sent a nice letter saying, no.

My own decisions descended from a mass of unsorted data stored somewhere pretty far back in the mind that interacted un-predictably with some kind of instantaneous romantic notion, in this case our grateful cows. At least some of Jean's decisions were made this way also. When she saw Bob's photo, she knew too.

The decision was easy for us but it was a terrible blow to the children. Jeanie was a student at Davidson College, and Ross was at Michigan State. Elizabeth, in high school, took the full brunt since she was still at home. Ross, always optimistic and easy, was diplo-matic. Jeanie felt betrayed. We had made the decision to sell things that they felt were as much theirs as ours. They had spent three quarters of their young lives on the place; Jean and I had invested only a quarter of ours. The children had the romantic notion that Jean and I could maintain the farm as it was without their presence - without their help - the romantic fiction that although they were gone that they were still here. Jean and I hurt, in our bones and joints, and we hurt for our children and their dreams.

187

After several weeks of bed rest failed to relieve my symptoms I went to surgery again. Another fragment of disk was found pressing on the nerve and was removed. My recovery was swift, the result was good, and the historic brick house went on the market along with the barns and a portion of the surrounding lands.

A bit later, Jean and I walked up through the Red Cow and the Hemlock and along the length of the White Gate pasture to the side of a rocky outcrop that rose above the rest of the field. Protected from the cold wind by the outcrop to the west and a steep hill to the north, this small patch of ground was usually the first to lose its snow during the spring. On this day as on most, the spot was friendly and inviting. We picked up a few weathered sticks that had been shed by the mature pines and we poked them into the ground, laying out the foundation for our future house. We canted the main axis just a bit west of south in order to maximize the solar gain from the incoming sun. We would keep the River Meadow, much of the woodland and determine which of the pastures would be sold with the house.

From a short distance down the field from here we could see our house and barns below laid out in a line along River Road - the blue of the river, and Vermont beyond. This was the same spot where eleven years earlier Mike and I had crouched in case the Squire should shoot. Now Jean and I stood up tall. The fences were straight and strong, the fields were full of clover, the barn roofs were in good repair. Several cows and some sheep grazed in the fields; there was laundry on the clothesline, firewood stacked for the winter. Our emotions were in our throats and were tempered only by excitement over the adventure we were soon to begin. Neither of us said a thing as we walked hand in hand down through the Red Cow and across the Home Meadow to the mud-room door.

We stepped onto the bellows-like floor. As our footsteps made the floor bounce, the door painted with Clef and Squire's names on the far side of the room rattled its latch as if to say hello. It was the same cheerful greeting that we had heard as we came through the door for the last 11 years. The rattle told us that the house had no problem with interlopers - those who were moving on. It was here to stay.

Epilogue

A few weeks ago, I found myself next to the logging crib, the place where the children had their fort. The tin roof is gone, of course. It has been fifty years since they put it on. I was struck by the diameter of the hemlock tree that had grown up in front of the crib - on the bare spot where we had walked so many times. It is 16 inches thick, tall and bushy. Soon the crib will be completely obscured by the tree and others nearby. This land, its abundant rainfall, freezing, thawing, and wonderful fertility is great at hiding the passing of man.

Time has also diminished the hurt the children experienced when we sold the house and barns. The children have by now learned that all parents disappoint their children at one time or another. The breaking away process is never without pain and I recognize the particularly strong attachment to place that was disrupted by our decision to sell. They have gone out into an ever more challenging world and their accomplishments have made me proud and happy. Some of the greatest pleasure I have at this stage of my life is to walk with them on the old familiar paths that we retained when we sold the house. On such walks artifacts left from our farming days provoke intense emotions and memories of past triumphs and failures.

The widowmaker has fallen and its stump is now hidden by new growth, but a careful observer could still use it as a landmark for a

190

buried object. The Hemlock Pasture is clear of the ground pine that inspired its name. Years of cutting, pulling, fertilizer, and a brush fire that escaped control one Palm Sunday took the starch out of that invader.

We placed a conservation easement on the River Meadow. More than a quarter mile of river frontage and meadow will stay undeveloped in perpetuity. Later we conserved a corridor leading from the River Meadow up past Hematology Falls, though the wooded section called the Territory and up to the property line of the Smith's property. This, too, awaits whatever geologic catastrophe signifies the end of perpetuity. We continued to own the River Meadow, the Red Cow, Hemlock, and White Gate Pastures, the High Pasture, and much of the woodland. Each year more people use the trails on the conserved land for hiking, cross country skiing or mountain biking and for horseback riding.

Don and Jennifer Cooke purchased the house, the Home Meadow, the Potato Meadow and the North Ring. The house looks better than ever. The stone wall that Dick Jenks built is now a feature of the lovely landscaping job that the new owners have done on that side of the house.

Bernard and Fran Tullar retired from farming and Bernard recently passed away. Fran just turned 100. We no longer make hay. Sometimes baled hay is delivered to our small new barn and then I toss bales up onto the stacks in the barn. The Tullar's herd got up to nearly 300 cows before milk prices plunged and they switched to producing organically certified milk. We are not called when their cows get out.

Granddaughter Chelsea cared for Jeanie's old pony, Curtsy. While reaching a remarkable age the pony developed increasingly shaggy hair and went nearly blind with cataracts. The pony continued her attempts to abuse other horses when she could see them until, in failing health, she was "put down". My fantasy of a cross-country walking trip leading Jeanie's pony, Carina, never came to pass. I have yet to find a pony that can lead as easily or one that I would rather put my pack on. She developed laminitis and after a long struggle she also had to be put down.

The Allis is still running, but needs another paint job. We sold the VAC and Ford 9N for more than we paid for them. (It is hard to go wrong investing in tractors.) For all I know the Super 66 baler is still working fine on some small New England farm. The Squire did not live out the year in which he sold the farm, dying of a ruptured aneurysm during a Florida winter. Clef, I'm sure, is gone also, but I don't know where or how.

George Lawson got tired of working on Walt Record's manure-encrusted dump truck and moved away, only to die early. Walt Record, before his death, conserved much of his land. His widow, Amy, moved to Vermont and operates a portable sawmill. I called on Dick Jenks in the hospital as he dealt with a cruel final illness. He was, as expected, as good a man as he could be in his situation.

The auctions at East Thetford Commission Sales are a thing of the past. Carleton Gray died suddenly, and the business was taken over by his son, Herb and family. Recently, as small farms have fallen by the wayside, the Grays gave up the Monday night commission sales. They continue to offer large consignment sales at regular intervals. These cannot replace the robust entertainment of the Monday night auctions, however.

George Evans, Clyde Grant, Lockwood Reed, Bunny Horton, Ed Blaisdell, Gordon Heard, Fred Hanchett, Glen Buzzell, Gerald Hewes, Bob MacDonnell, Ralph Balch and other old friends have passed on. I should have spent more time with them.

Jean continued to raise sheep selecting them for the quality of their fleece. She sold wool to hand spinners. Her spinning wheels hummed for hours as her knitting projects grew to a fine art. From the small barn at our new house, lambs burst forth each spring onto the White Gate Meadow to remind us of the time when our farm life was out of control. We went away for several months to Alaska. We had acquired an Italian sheep guard dog, a Maremma, to protect the flock from predators while we were gone. Later, when the dog became vicious, she was replaced with a gentle and conscientious guard llama. Throughout this happy period in the sheep business none were lost.

When Jean became seriously ill in 2001 it was clear that our decision to simplify our lives had been a good one. In the next three and a half years we were able to have lots of happy times to offset the occasional hard knocks delivered by the disease and its treatment. During this period, Jean quietly told us of a time years ago when, unknown to her family, she had loosed her horse for furious gallops along our woods trails or had taken solo trips through challenging country. As the children and I came to understand this hidden aspect of her personality, we found the source of the courage with which she faced her cruel disease. Jean and I slept in the same bed until the night she died in 2005. Now as I walk some of trails alone, I imagine a thunder of hooves and my bride of years ago flashing by.

In 2006 I met Helen Whyte, a woman with a background in planning and development, whose attachment to the remote shore of Manitoulin Island in Ontario is similar to that of mine to Riverbridge Farm. We were married in 2008 and intend to experience as much of the world we can as a pair.

Often I walk along the River Meadow near where Ross's brook empties into the Connecticut River. From there I can look up to the old brick house that sheltered us for those happy years. It is in good shape now, well taken care of. Not long ago when I walked there through a stubble of harvested corn, a product of that rich earth, plenty of rain, and warm sun, I encountered a flock of crows. They spoke raucously to each other as they picked the kernels left behind by the chopper. As I grew close they flew up and away to leave me on my own. In the stillness now, I watched the river passing quietly, deep and steady, on its way to the sea.

Acknowledgements

Many friends and neighbors have assisted me and my family as we inflicted our farming adventure upon them. Their help has been unstinting and essential to any successes that we have had. They have found themselves witnesses to the events recorded here. The members of the Wednesday Morning Memoir Group have offered advice and help as I have read portions the manuscript to the group. My neighbors Mike and Jean Smith have been active players in these events and without their participation in the start of the story there would not have been one. Their children and extended family have been keen stewards of the Stone Farmhouse Trails network that includes the entirety the Riverbridge trail system. Don and Jen Cooke who purchased our house and portions of the farmland described in this tale have been most generous in granting continued access to it. Jen's improvements have ensured the preservation of the house and enhanced its historic value.

Kate Emlen, a one time neighbor, whose posters have brightened the walls of our new house, made the sketch map of the farm for this book. David Caffry let me photograph his rooster for the cover, and Richard Granger fine-tuned the rooster image for the book. Bill Hudenco flew his drone over the farm for the photo in chapter one. One cannot live the kind of life we have enjoyed for these many years, except in a community that supports tolerance, values achievement, and preserves history. Thank you Lyme.

Jean Geary McIntyre who spread her arms to support and lead her three growing children throughout the years of this story kept the saga going and infused with happiness. My marriage to her was such happy one that after her death it encouraged me to try again. Helping me to edit this book, Helen Whyte has learned enough of the history of Riverbridge Farm to have a full membership in it too. To the children I offer more thanks than can be measured, and I apologize for invading their lives to let others know about childhood on a farm. Without them the story would not exist.

About the Author

During the eleven years covered in this memoir, the author served initially as Chief of the Section of Hematology and later as Director of the Norris Cotton Cancer Center at the Dartmouth-Hitchcock Medical Center. A graduate of Dartmouth College and Harvard Medical School, Dr. McIntyre finished his career as the elected Chairman of the Cancer and Leukemia Group B, a National Cancer Institute supported Clinical Trials Group. Many friends, patients, classmates and former students have endowed the O. Ross McIntyre Professorship at the Dartmouth-Geisel School of Medicine. O. Ross has numerous publications in scientific journals and medical textbooks. His memoir of canoe tripping, *Paddle Beads*, GraybooksPublishers, appeared in 2010.

www.ingramcontent.com/pod-product-compliance
Lightning Source LLC
Chambersburg PA
CBHW060513130626
46553CB00002B/478